Praise for *I'm OK—You're OK*

"Harris has stripped away the technical language of psychoanalysis and presented with lucid logic a way to self-understanding and change." —*Los Angeles Times*

"When an idea finds its time and voice it takes on force. Transactional Analysis is the idea. Now is the time. *I'm OK—You're OK* is the voice. With a little bit of luck and a lot of people trying to screw their heads on straight . . . *I'm OK—You're OK* may make it up there right next to the Holy Bible or maybe even *The Better Homes and Gardens Cookbook*." —*Life*

"The book reads easily and interestingly for the unlettered in behavior science. . . . The practicing therapist cannot help but benefit immeasurably, especially if his inclinations are toward family or group therapy." —*Choice*

"It's easy to relate to *I'm OK—You're OK*. Again and again the reader will find his own predicament staring up at him from the page, along with its message, 'If you don't like the trap, you can change it.' " —*Cleveland Press*

"[Readers] who roam these optimistic pages with their allusions to and discussions of Freud, Wilder Penfield, Elton Trueblood, Eric Berne, Bishop Pike, Teilhard de Chardin, and many other great and not-so-great experts may well make the book a bestseller." —*Library Journal*

"I am grateful to Dr. Harris for doing a job that needed doing. In this book he has clarified the principles of Transactional Analysis with cogent, easily understood examples and has related them to broader considerations, including ethics, in a thoughtful and skillful way." —Eric Berne, M.D.

Olan Mills

About the Author

The late THOMAS A. HARRIS, a graduate of Temple University Medical School and a career Navy psychiatrist, rose to become the Chief of Psychiatry in the Navy Department after surviving his first at-sea mission, the Pearl Harbor attack. Subsequent to his retirement as commander, Dr. Harris was a university professor and directed child guidance clinics, mental hospitals, youth correctional facilities, and a maximum security prison.

He entered private practice in Sacramento in 1956, founded an institute, and was a director of the International Transactional Analysis Association. Dr. Harris studied with Dr. Eric Berne, who conceived of Transactional Analysis, on which *I'm OK—You're OK* is based. Dr. Harris's earlier mentors included Dr. Harry Stack Sullivan and Dr. Frieda Fromm-Reichmann of the Washington-Baltimore Psychoanalytic Institute.

Life magazine called Dr. Harris "chatty and grandfatherly" and documented his gratitude toward his wife, Amy Harris, for her "skill in taking his ideas and making them readable."

I'm OK—
You're OK

Thomas A. Harris, M.D.

I'm OK –
You're OK

Quill
An Imprint of HarperCollins*Publishers*

Passages from *The Shaking of the Foundation* (copyright © 1948) and *The New Being* (copyright © 1955) by Paul Tillich, are used by permission of Charles Scribner's Sons.

Passages from *General Philosophy* by D. Elton Trueblood (copyright © 1963) are reprinted by permission of Harper & Row, Publishers.

First Quill edition published 2004.

Designed by Nancy B. Field

Library of Congress Cataloging-in-Publication Data is available upon request.
ISBN 0-06-072427-7

04 05 06 07 08 JTC/RRD 10 9 8 7 6 5 4 3 2 1

To Amy

my collaborator

my philosopher

my tranquilizer

my joy

my wife

Contents

Illustrations

A Note on the Publishing History

Since its initial hardcover publication in 1969 to widespread enthusiasm, this insightful classic of popular psychology has been a bestselling record-breaker—with more than 15 million copies in the hands of readers to date. For Harris, being OK with ourselves and with others is the way to happiness, personal satisfaction, and good relationships. The writing and publication of this book was an important step in furthering Dr. Harris's mission to change the world one person at a time.

Author's Note

It is important that this book be read from front to back. Were later chapters read before the first chapters, which define the method and vocabulary of Transactional Analysis, the reader not only would miss the full significance of the later chapters but would assuredly make erroneous conclusions.

Chapters 2 and 3 are particularly essential to the understanding of all that follows. For readers who have an irresistible back-to-front reading urge, I wish to emphasize that five words which appear throughout the book have specific meanings different from their usual meanings. They are "Parent," "Adult," "Child," "OK," and "games."

Preface

In recent years there have been many reports of a growing impatience with psychiatry, with its seeming foreverness, its high cost, its debatable results, and its vague, esoteric terms. To many people it is like a blind man in a dark room looking for a black cat that isn't there. The magazines and mental-health associations say psychiatric treatment is a good thing, but what it is or what it accomplishes has not been made clear. Although hundreds of thousands of words about psychiatry are consumed by the public yearly, there has been little convincing data to help a person in need of treatment overcome the cartoon image of psychiatrists and their mystical couches.

Impatience has been expressed with increasing concern not only by patients and the general public but by psychiatrists as well. I am one of these psychiatrists. This book is the product of a search to find answers for people who are looking for hard facts in answer to their questions about how the mind operates, why we do what we do, and how we can stop doing what we do if we wish. The answer lies in what I feel is one of the most promising breakthroughs in psychiatry in many years. It is called Transactional Analysis. It has given hope to people who have become discouraged by the vagueness of many of the traditional types of psychotherapy. It has given a new answer to people who want to change rather than to adjust, to people who want transformation rather than conformation. It is realistic in that it confronts the patient with the fact that he is responsible for what happens in the future no matter what has happened in the past. Moreover, it is enabling persons to change, to establish self-control and self-direction, and to discover the reality of a freedom of choice.

For the development of this method we are pre-eminently indebted to Dr. Eric Berne, who, in developing the concept of Transactional Analysis, has created a unified system of individual and social psychiatry that is comprehensive at the theoretical level and effective at the applied level. It has been my privilege to study with Berne for the past ten years and to share the discussions of the advanced seminar in San Francisco which he conducts.

I first became acquainted with Berne's new method of treatment through a paper that he presented at the Western Regional Meeting of the American Group Psychotherapy Association in Los Angeles in November 1957. It was entitled "Transactional Analysis: A New and Effective Method of Group Therapy." I was convinced that this was not "just another paper," but indeed a blueprint of the mind, which no one had constructed before, along with a precision vocabulary, which anybody could understand, to identify the parts of the blueprint. This vocabulary has made it possible for two people to talk about behavior and know what is meant.

One difficulty with many psychoanalytic words is that they do not have the same meanings for everybody. The word *ego,* for instance, means many things to many people. Freud had an elaborate definition, as has nearly every psychoanalyst since his time; but these long, complicated constructions are not particularly helpful to a patient who is trying to understand why he can never hold a job, particularly if one of his problems is that he cannot read well enough to follow instructions. There is not even agreement by theoreticians as to what ego means. Vague meanings and complicated theories have inhibited more than helped the treatment process. Herman Melville observed that "a man of true science uses but few hard words, and those only when none other will answer his purpose; whereas the smatterer in science . . . thinks that by mouthing hard words he understands hard things." The vocabulary of Transactional Analysis is the precision tool of treatment because, in a language anyone can understand, it identifies things that really are, the reality of experiences that really happened in the lives of people who really existed.

Also the method, which is particularly suited to the treatment of people in groups, points to an answer to the great disparity between the need for treatment and the trained people available to do the work. During the past twenty-five years, beginning with particular intensity in the years immediately following World War II, the popularity of psychiatry would seem to have created expectancies far beyond our capacity to fulfill them. Continual outpourings of psychological literature, whether printed in psychiatric journals or the *Reader's Digest*, have increased this expectancy yearly, but the chasm between this and cure seems to have widened. The question has always been how to get Freud off the couch and to the masses.

The challenge to psychiatry to meet this need was expressed by Mike Gorman, the Executive Director of the National Committee Against Mental Illness, in an address to the annual meeting of the American Psychiatric Association in New York in May 1965:

> As you have escalated from a small cell of some 3,000 psychiatrists in 1945 to a large specialty organization with 14,000 members in 1965, you have of necessity been increasingly drawn into participation in the major issues of our time. You can no longer hide in the discomfort of your private office, appropriately fitted with an overstuffed conch and a picture of Freud visiting Worcester, Massachusetts, in 1909.
>
> I submit that *psychiatry must develop a "public" language, decontaminated of technical jargon and suited to the discussion*

of universal problems of our society. I realize that this is a very difficult task; it means taking leave of the comfortable, secure, and protected words of the profession and adjusting to the much breezier dialogue of the open tribunal. As difficult as this task is, it must be done if psychiatry is to be heard in the civic halls of our nation.

I am heartened by the recent writings of a number of young psychiatrists which demonstrate a healthy aversion to spending an entire professional life treating ten to twenty patients a year.

The comment of the psychiatrist Dr. Melvin Sabshin is typical: "One simple question is whether or not psychiatry can accomplish these new functions or roles by utilizing its traditional skills, its standard methodology, and its current practices. My own answer to the question is no. I believe these do not provide an adequate basis for new functions and configurations."

Psychiatry must face up to the fact that it cannot begin to meet the demands for psychological and social help from the poor, the underachieving in our schools, the frustrated among our blue collar workers, the claustrophobic residents in our crowded cities, and so on almost *ad infinitum.*

Many of its most thoughtful leaders are giving increasing thought to the new role which psychiatry must play during the next several decades, in not only broadening its own parochial training, but in joining with other behavioral disciplines on an equal footing in establishing training programs for the thousands upon thousands of new mental health workers we will need if we are to achieve the goals which President Kennedy proclaimed in his historic 1963 mental health message.*

Training programs of thousands of mental-health workers in a " 'public' language, decontaminated of technical jargon and suited to the discussion of universal problems in our society" is being made possible today by Transactional Analysis. More than 1,000 professionals have been trained in this method in the State of Cal-

*M. Gorman, "Psychiatry and Public Policy," *The American Journal of Psychiatry,* Vol. 122, No. 1 (July 1965).

ifornia, and this training is spreading rapidly to other parts of the country and to foreign countries. About one-half of these professionals are psychiatrists; the other half includes medical doctors of other specialties (obstetrics, pediatrics, internal medicine, general practice), psychologists, social workers, probation officers, nurses, teachers, personnel managers, clergymen, and judges. Transactional Analysis is now being used in group treatment in many of California's state hospitals, prisons, and Youth Authority institutions. It is used by increasing numbers of therapists in marital counseling, treatment of adolescents and preadolescents, pastoral counseling, and family-centered obstetrical care, and in at least one institution for the mentally retarded, Laurel Hills of Sacramento.

A central reason why Transactional Analysis offers such promise for filling the gap between need for and supply of treatment is that it works at its best in groups. It is a teaching and learning device rather than a confessional or an archeological exploration of the psychic cellars. In my private practice of psychiatry this has made possible the treatment of four times as many patients as before. During the past twenty-five years in my work as a psychiatrist—in the treatment of patients and in the administration of large institutional programs—nothing has excited me so much as what is happening today in my practice. One of the most significant contributions of Transactional Analysis is that it has given patients a tool they can use. The purpose of this book is to define this tool. Anybody can use it. People do not have to be "sick" to benefit from it.

It *is* a profoundly rewarding experience to see people begin to change from the first treatment hour, get well, grow, and move out of the tyranny of the past. We base our even greater hope on the affirmation that what has been can be again. If the relationship between two people can be made creative, fulfilling, and free of fear, then it follows that this can work for two relationships, or three or one hundred or, we are convinced, for relationships that affect entire social groups, even nations. The problems of the world—and they are

chronicled daily in headlines of violence and despair—essentially
are the problems of individuals. If individuals can change, the
course of the world can change. This is a hope worth sustaining.

I wish to thank a number of people for their support of, and contribution to, the effort involved in writing this book. Mostly I owe
the reality of this book to my wife Amy, whose writing skill and
phenomenal thought processes have put into this final form the
content of my lectures, research, past writings, observations, and
formulations, many of which we worked out together. Evidences of
her philosophical, theological, and literary researches are sprinkled
throughout the book, and the chapter on moral values is her original contribution. Also I express appreciation to my secretaries Bev-
erly Fleming and Connie Drewry, who prepared the typescript and
study copies of the manuscript; to Alice Billings, Merrill Heidig,
Jean Lee, Margery Marshall, and Jan Root for their valuable assistance; to my children for their delightful contribution;

To my colleagues who joined me in founding the Institute for
Transactional Analysis: Dr. Gordon Haiberg, Dr. Erwin Eichhorn,
Dr. Bruce Marshall, Rev. J. Weaver Hess, and John R. Saldine; to
the directors who joined us as the Institute Board expanded: Dr.
David Applegate, Laverne Crites, Dr. Donis Eichhorn, Dr. Ronald
Fong, Dr. Alvyn Freed, David Hill, Dr. Dennis Marks, Larry Mart,
Dr. John Mitchell, Richard Nicholson, Rev. Russell Osnes, Dr.
Warren Prentice, Berton Root, Barry Rumbles, Frank Summers,
Rev. Ira Tanner, Leroy Wolter, and Dr. Z. O. Young;

To the late Rev. Dr. Robert R. Ferguson, Senior Pastor of Fremont Presbyterian Church of Sacramento and consultant in field education at Princeton Theological Seminary; to Dr. John M. Campbell,
Chairman of the Department of Anthropology at the University of
New Mexico; to James J. Brown of the *Sacramento Bee;* to Eric Bjork
for his wisdom and generous commentary; to Dr. Ford Lewis, Minister of the First Unitarian Society of Sacramento, whose devotion to
truth and compassion has been a rich source of encouragement;

To Dr. Elton Trueblood, Professor of Philosophy at Earlham College, for the significant new data he made available to me; to Bishop James Pike, Resident Theologian at the Center for the Study of Democratic Institutions at Santa Barbara, for his contagious enthusiasm and generous assistance; to two special persons who provided years of training and stimulation, Dr. Freida Fromm-Reichmann and Dr. Harry Stack Sullivan, in whose tutelage I first heard the term "interpersonal transactions";

And finally to my patients, whose creative and emancipated thinking has provided much of the content of this book. It is at their request that I have written it.

T.A.H.

Sacramento, California
June 1968

I'm OK–
You're OK

Freud, Penfield,
and Berne

I contradict myself. I am large. I contain multitudes.

—Walt Whitman

Throughout history one impression of human nature has been consistent: that man has a multiple nature. Most often it has been expressed as a dual nature. It has been expressed mythologically, philosophically, and religiously. Always it has been seen as a conflict: the conflict between good and evil, the lower nature and the higher nature, the inner man and the outer man. "There are times," said Somerset Maugham, "when I look over the various parts of my character with perplexity. I recognize that I am made up of several persons and that the person that at the moment has the upper hand will inevitably give place to another. But which is the real one? All of them or none?"

That man can aspire to and achieve goodness is evident through all of history, however that goodness may be understood. Moses saw goodness supremely as justice, Plato essentially as wisdom, and Jesus centrally as love; yet they all agreed that virtue, however understood, was consistently undermined by something in human nature which was at war with something else. But what were these somethings?

When Sigmund Freud appeared on the scene in the early twentieth century, the enigma was subjected to a new probe, the discipline of scientific inquiry. Freud's fundamental contribution was his theory that the warring factions existed in the unconscious. Tentative names were given to the combatants: the Superego became thought of as the restrictive, controlling force over the Id (instinctual drives), with the Ego as a referee operating out of "enlightened self-interest."

We are deeply indebted to Freud for his painstaking and pioneering efforts to establish the theoretical foundation upon which we build today. Through the years scholars and clinicians have elaborated, systematized, and added to his theories. Yet the "persons within" have remained elusive, and it seems that the hundreds of volumes which collect dust and the annotations of psychoanalytic thinkers have not provided adequate answers to the persons they are written about.

I stood in the lobby of a theater at the end of the showing of the motion picture *Who's Afraid of Virginia Woolf?* and listened to a number of comments by people who had just seen the picture: "I'm exhausted!" "And I come to movies to get away from home." "Why do they want to show something like that?" "I didn't get it; I guess you have to be a psychologist." I got the impression that many of these people left the theater wondering what was *really* going on, sure there must have been a message, but unable to find anything relevant to them or liberating in terms of how to end "fun and games" in their own lives.

We are dutifully impressed by formulations such as Freud's definition of psychoanalysis as a "dynamic conception which reduces mental life to an interplay of reciprocally urging and checking forces." Such a definition and its countless elaborations may be useful to "the professionals," but how useful are these formulations to people who hurt? George and Martha in Edward Albee's play used red-hot, gutsy, four-letter words that were precise and to the point. The question is, As therapists can we speak with George and Martha as precisely and pointedly about *why* they act as they do and hurt as they do? Can what we say be not only true but also helpful, because we are understood? "Speak English! I can't understand a word you're saying," is not an uncommonly held attitude toward persons who claim to be experts in the psychological fields. Restating esoteric psychoanalytic ideas in even more esoteric terms does not reach people where they live. As a consequence the reflections of ordinary folk are often expressed in pitiful redundancies and in superficial conversations with such summary comments as, "Well, isn't that always the way?" with no understanding of how it can be different.

In a sense, one of the estranging factors of the present day is the lag between specialization and communication, which continues to widen the gulf between specialists and nonspecialists. Space belongs to the astronauts, understanding human behavior belongs to the psychologists and psychiatrists, legislation belongs to the congressmen, and whether or not we should have a baby belongs to the theologians. This is an understandable development; yet the problems of nonunderstanding and noncommunication are so great that means must be devised whereby language can keep up with the developments of research.

In the field of mathematics an answer to this dilemma was attempted in the development of the "new mathematics," which has been taught in elementary schools throughout the country. The new mathematics is not so much a new form of computation as of communication of mathematical ideas, answering questions not only of *what*, but also of *why*, so that the excitement of going to the moon or using a computer will not remain exclusively in the realm of scientists but can also exist in comprehensible form for the student. The science of mathematics is not new, but the way it is talked about is new. We would find ourselves handicapped if we were still to use the Babylonian, Mayan, Egyptian, or Roman number systems. The desire to use mathematics creatively brought about new ways of systematizing numbering concepts. The new mathematics of today has continued this creative growth. We recognize and appreciate the creative thinking which the earlier systems represented, but we do not encumber today's work with those sometimes less effective methods.

This is my position with regard to Transactional Analysis. I respect the devoted effort of the psychoanalytic theorists of the past. What I hope to demonstrate in this book is a new way to state old ideas and a clear way to present new ones, not as an inimical or deprecating assault on the work of the past, but rather as a means of meeting the undeniable evidence that the old methods do not seem to be working very well.

Once, an old farmer, tinkering with a rusty harrow on a country road, was approached by an earnest young man from the University Extension Service who was making farm-to-farm calls for the purpose of selling a new manual on soil conservation and new farming techniques. After a polite and polished speech the young man asked the farmer if he would like to buy this new book, to which the old man replied,

"Son, I don't farm half as good as I know how already."

The purpose of this book is not only the presentation of new data but also an answer to the question of why people do not live as good as they know how already. They may know that the experts have had a lot to say about human behavior, but this knowledge does not seem to have the slightest effect on their hangover, their splintering marriage, or their cranky children. They may turn to "Dear Abby" for advice or find themselves delightfully portrayed in "Peanuts," but is there anything both profound and simple related to the *dynamics of behavior* which will help them find new answers to old problems? Is there any information available which is both true and helpful?

Our search for answers has until recent years been limited by the fact that we have known relatively little about how the human brain stores memory and how this memory is evoked to produce the tyranny—as well as the treasure—of the past in current living.

The Brain Surgeon with the Probe

Any hypothesis must depend for its verification on observable evidence. Until recently there has been little evidence about how the brain functions in cognition, precisely how and which of the billions of cells within the brain store memory. How much memory is retained? Can it disappear? Is memory generalized or specific? Why are some memories more available for recall than others?

One noted explorer in this field is Dr. Wilder Penfield, a neurosurgeon from McGill University in Montreal, who in 1951 began to produce exciting evidence to confirm and modify theoretical concepts which had been formulated in answer to these questions.* During the course of brain surgery, in treating patients suffering from focal epilepsy, Penfield conducted a series of experiments during which he touched the temporal cortex of the brain of the patient with a weak electric current transmitted through a galvanic probe. His observations of the responses to these stimulations were accumulated over a period of several years. In each case the patient under local anesthesia was fully conscious during the exploration of the cerebral cortex and was able to talk with Penfield. In the course of these experiments he heard some amazing things.

(Inasmuch as this book is meant to be a practical guide to Transactional Analysis and not a technical scientific treatise, I wish to clarify that the following material from Penfield's research—the only material in this book which might be seen as technical—is included in the first chapter because I believe it is essential to the es-

*W. Penfield, "Memory Mechanisms," *A.M.A. Archives of Neurology and Psychiatry* 67(1952): 178–198, with discussion by L. S. Kubie et al. Quotations from Penfield and Kubie later in this chapter are from the same source.

tablishment of the scientific basis of all that follows. The evidence seems to indicate that everything which has been in our conscious awareness is recorded in detail and stored in the brain and is capable of being "played back" in the present. The following material may warrant more than a single reading for a full appreciation of the implications of Penfield's findings.)

Penfield found that the stimulating electrode could force recollections clearly derived from the patient's memory. Penfield reported, "The psychical experience, thus produced, stops when the electrode is withdrawn and may repeat itself when the electrode is reapplied." He gave the following examples:

> First is the case of S.B. Stimulation at Point 19 in the first convolution of the right temporal lobe caused him to say: "There was a piano there and someone was playing. I could hear the song, you know." When the point was stimulated again without warning, he said: "Someone speaking to another," and he mentioned a name, but I could not understand it . . . it was just like a dream. The point was stimulated a third time, also without warning. He then observed spontaneously, "Yes, *Oh Marie, Oh Marie!*—Someone is singing it." When the point was stimulated a fourth time, he heard the same song and explained that it was the theme song of a certain radio program.
>
> When Point 16 was stimulated, he said, while the electrode was being held in place, "Something brings back a memory. I can see Seven-Up Bottling Company . . . Harrison Bakery." He was then warned that he was being stimulated, but the electrode was not applied. He replied, "Nothing."
>
> When, in another case, that of D.F., a point on the superior surface of the right temporal lobe was stimulated within the fissure of Sylvius, the patient heard a specific popular song being played as though by an orchestra. Repeated stimulations reproduced the same music. While the electrode was kept in place, she hummed the tune, chorus, and verse, thus accompanying the music she heard.
>
> The patient, L.G., was caused to experience "something,"

he said, that had happened to him before. Stimulation at another temporal point caused him to see a man and a dog walking along a road near his home in the country. Another woman heard a voice which she did not quite understand when the first temporal convolution was stimulated initially. When the electrode was reapplied to approximately the same point, she heard a voice distinctly calling, "Jimmie, Jimmie"—Jimmie was the nickname of the young husband to whom she had been married recently.

One of Penfield's significant conclusions was that the electrode evoked a single recollection, not a mixture of memories or a generalization.

Another of his conclusions was that the response to the electrode was involuntary:

Under the compelling influence of the probe a familiar experience appeared in the patient's consciousness whether he desired to focus his attention upon it or not. A song went through his mind, probably as he had heard it on a certain occasion: he found himself a part of a specific situation that progressed and evolved just as the original situation did. It was, to him, the act of a familiar play, and he was himself both an actor and the audience.

Perhaps the most significant discovery was that not only past events are recorded in detail but also the feelings that were associated with those events. An event and the feeling which was produced by the event are inextricably locked together in the brain so that one cannot be evoked without the other. Penfield reported:

The subject feels again the emotion which the situation originally produced in him, and he is aware of the same interpretations, true or false, which he himself gave to the experience in the first place. Thus, evoked recollection is not the exact photographic or phonographic reproduction of past scenes or events.

It is reproduction of what the patient saw and heard and felt and understood.

Recollections are evoked by the stimuli of day-to-day experience in much the same way that they were evoked artificially by Penfield's probe. In either case the evoked recollection can be more accurately described as a *reliving* than a recalling. In response to a stimulus a person is momentarily displaced into the past. *I am there!* This reality may last only a fraction of a second, or it may last many days. Following the experience a person may then consciously *remember* he was there. The sequence in involuntary recollections is: (1) *reliving* (spontaneous, involuntary feeling), and (2) *remembering* (conscious, voluntary thinking about the past event thus relived). Much of what we relive we cannot remember!

The following reports of two patients illustrate the way in which stimulations in the present evoke past feelings.

A forty-year-old female patient reported she was walking down the street one morning and, as she passed a music store, she heard a strain of music that produced an overwhelming melancholy. She felt herself in the grip of a sadness she could not understand, the intensity of which was "almost unbearable." Nothing in her conscious thought could explain this. After she described the feeling to me, I asked her if there was anything in her early life that this song reminded her of. She said she could not make any connection between the song and her sadness. Later in the week she phoned to tell me that, as she continued to hum the song over and over, she suddenly had a flash of recollection in which she "saw her mother sitting at the piano and heard her playing this song." The mother had died when the patient was five years old. At that time the mother's death had produced a severe depression, which had persisted over an extended period of time, despite all the efforts of the family to help her transfer her affection to an aunt who had assumed the mother role. She had never recalled hearing this song or remembering her mother's playing it until the day she walked by the music store. I asked her if the recall of this early memory had relieved her of the depression. She said it had changed the nature of her feelings; there was still a melancholy feeling in recalling the death of her mother, but it was not the initial overwhelming despair she felt at first. It would seem she was now consciously remembering a feeling which initially was the *reliving* of a feeling. In the second instance, she remembered how it was to feel that way; but in the first instance, the feeling was precisely the *same* feeling which was recorded when her mother died. She was at that moment five years old.

Good feelings are evoked in much the same way. We are all aware of how an odor, a sound, or a fleeting glimpse can produce an ineffable joy, sometimes so momentary it almost goes unnoticed. Unless we put our minds to it, we cannot remember where we had experienced the smell, sound, or sight before. But the *feeling* is real.

Another patient reported this incident. He was walking along L Street by Sacramento's Capitol Park and, upon smelling the odor of lime and sulphur, generally thought to be putrid, being used as a spray for the trees, he was aware of a glorious carefree feeling of joy. Uncovering the original situation was easier for him since the feeling was a good one. This was the kind of spray that had been used in the early spring in his father's apple orchard and, for the patient as a little boy, this smell was synchronous with the coming of spring, the "greening" of the trees, and all the joys experienced by a little boy emancipated to the outdoors after the long winter. As in the case of the first patient, the conscious remembering of the feeling was slightly different from the burst of the original feeling that he experienced. He could not quite recapture the glorious, spontaneous transference into the past as he did for that fleeting moment. It was as if he now had *a feeling about his feeling* rather than the feeling itself.

This illustrates another of Penfield's conclusions: the memory record continues intact even after the subject's ability to recall it disappears:

Recollection evoked from the temporal cortex retains the detailed character of the original experience. When it is thus introduced into the patient's consciousness, the experience seems to be in the present, possibly because it forces itself so irresistibly upon his attention. Only when it is over can he recognize it as a vivid memory of the past.

Another conclusion we may make from these findings is that the brain functions as a high-fidelity recorder, putting on tape, as it were, every experience from the time of birth, possibly even before birth. (The process of information storage in the brain is undoubtedly a chemical process, involving data reduction and coding, which is not fully understood. Perhaps oversimple, the tape recorder analogy nevertheless has proved useful in explaining the memory process. The important point is that, however the recording is done, the playback is high fidelity.)

Whenever a normal person is paying conscious attention to something [says Penfield], he simultaneously is recording it in the temporal cortex of each hemisphere.

These recordings are in sequence and continuous.

When the electrode is applied to the memory cortex it may produce a picture, but the picture is not usually static. It changes, as it did when it was originally seen and the subject perhaps altered the direction of his gaze. It follows the originally observed events of succeeding seconds or minutes. The song produced by cortical stimulation progresses slowly, from one phrase to another and from verse to chorus.

Penfield further concludes that the thread of continuity in evoked recollections seems to be *time*. The original pattern was laid down in temporal succession.

The thread of temporal succession seems to link the elements of evoked recollection together. It also appears that only those sensory elements to which the individual was paying attention are recorded, not all the sensory impulses which are forever bombarding the central nervous system.

The evoking of complicated memory sequences makes it seem plausible that each of the memories we can recall has a separate neurone pathway.

Particularly significant to our understanding of how the past influences the present is the observation that the temporal cortex is obviously utilized in the interpretation of current experience.

Illusions . . . may be produced by stimulation of the temporal cortex . . . and the disturbance produced is one of judgment in regard to present experience—a judgment that the experience is familiar, or strange, or absurd; that distances and sizes are altered, and even that the present situation is terrifying.

These are illusions of perception, and a consideration of them leads one to believe that a *new experience is somehow immediately classified together with records of former similar experience so that judgment of differences and similarities is possible.* For example, after a period of time it may be difficult for a man to conjure up an accurate, detailed memory of an old friend as he appeared years ago, and yet when the friend is met, however unexpectedly, it is possible to perceive at once the change that time has wrought. One knows it all too well—new lines in his face, change in hair, stoop of shoulder. [Italics mine.]

Penfield concludes:

The demonstration of the existence of cortical "patterns" that preserve the detail of current experience, as though in a library of many volumes, is one of the first steps toward a physiology of the mind. The nature of the pattern, the mechanism of its formation, the mechanism of its subsequent utilization, and the in-

tegrative processes that form the substratum of consciousness—
these will one day be translated into physiological formulas.

Dr. Lawrence S. Kubie of Baltimore, one of the nation's prominent psychoanalysts who was among the discussants of Penfield's
paper, said, at the conclusion of the presentation:

> I am profoundly grateful for this opportunity to discuss
> Doctor Penfield's paper . . . because of the enormous stimula-
> tion which the paper itself has given to my imagination. In-
> deed it has kept me in a state of ferment for the last two weeks,
> watching pieces of a jigsaw puzzle fit into place and a picture
> emerge to throw some light on some of the work which I have
> been doing in recent years. I can sense the shades of Harvey
> Cushing and Sigmund Freud shaking hands over this long-
> deferred meeting between psychoanalysis and modern neuro-
> surgery through the experimental work which Doctor Penfield
> has reported.

In summary we may conclude:

1. The brain functions as a high-fidelity tape recorder.
2. The feelings which were associated with past experiences
also are recorded and are *inextricably locked* to those experiences.
3. Persons can exist in two states at the same time. The patient
knew he was on the operating table talking with Penfield; he
equally knew he was seeing the "Seven-Up Bottling Company . . .
and Harrison Bakery." He was *dual* in that he was at the same time
in the experience and *outside* of it, observing it.
4. These recorded experiences and *feelings associated with
them* are available for replay today in as vivid a form as when they
happened and provide much of the data which determines the nature of today's transactions. These experiences not only can be recalled but also relived. *I not only remember how I felt. I feel the same
way now.*

Penfield's experiments demonstrate that the memory function, which is most often thought of in psychological terms, is biological also. We are not able to answer the age-old question of how the mind is attached to the body. It is pertinent, however, to refer to the rapid progress being made in the field of genetic research as to how heredity is programmed within the RNA molecule. Sweden's Dr. Holgar Hyden has reflected:

> The capacity to recall the past to consciousness can certainly be expected to reside in a primary mechanism of general biological validity. A firm link to the genetic mechanism is important, and in this respect especially, the RNA molecule, with its many possibilities, would fulfill many requirements.*

The observable evidence produced by these biological studies supports and helps to explain the observable evidence in human behavior. How do we apply the scientific method to behavior in such a way that our findings constitute as accurate and as useful a body of "knowns" as Penfield's findings?

A Basic Scientific Unit: The Transaction

One of the reasons for the criticism that the psychotherapeutic sciences are unscientific, and for much of the disagreement evident in this field, is that there has been no basic unit for study and observation. It is the same kind of difficulty as that which confronted physicists before the molecular theory and physicians before the discovery of bacteria.

Eric Berne, the originator of Transactional Analysis, has isolated and defined this basic scientific unit:

> The unit of social intercourse is called a transaction. If two or more people encounter each other . . . sooner or later one of

*H. Hyden, "The Biochemical Aspects of Brain Activity," in S. M. Farber and R. Wilson, eds., *Control of the Mind* (New York: McGraw-Hill, 1961), p. 33.

them will speak, or give some other indication of acknowledging the presence of the others. This is called the *transactional stimulus*. Another person will then say or do something which is in some way related to the stimulus, and that is called the *transactional response*.*

Transactional Analysis is the method of examining this one transaction wherein "I do something to you and you do something back" and determining which part of the multiple-natured individual is "coming on." In the next chapter, "Parent, Adult, and Child," the three parts of this multiple nature are identified and described.

Transactional Analysis also is the method of systematizing the information derived from analyzing these transactions in words which have the same meaning, by definition, for everyone who is using them. This language is clearly one of the most important developments of the system. Agreement on the meanings of words plus agreement on what to examine are the two keys which have unlocked the door to the "mysteries of why people do as they do." This is no small accomplishment.

In February, 1960, I had the opportunity of hearing a fascinating, day-long dissertation by Dr. Timothy Leary, who had then just joined the Department of Social Relations at Harvard University. He spoke to the staff of DeWitt State Hospital in Auburn, California, where I was Director of Professional Education. Despite the controversial responses he now evokes by his devotion to the use of drugs in the pursuit of psychedelic experience, I wish to use some of his comments here, inasmuch as they express the problem dramatically and may explain what he called his own "zigzag course of sequential disillusionment." He stated that one of his greatest frustrations as a psychotherapist was the inability to discover a way to standardize language and observation about human behavior:[†]

*E. Berne, *Games People Play* (New York: Grove Press, 1964), p. 29.
†T. Leary, address, Dewitt State Hospital, Auburn, California, Feb. 23, 1960. Quotations from Leary later in the chapter are from the same address.

I would like to share with you some of the historical back-ground of my immobilization as a psychological scientist. As I look back I can see that there were three stages of my own ignorance. The first, which was by far the most happy, you could call the stage of innocent ignorance when I was possessed with the notion that there were some secrets of human nature, there were some laws and regularities, some cause and effect relationships, and that through study, through experiences, through reading, some day I would share these secrets and be able to apply my knowledge of these regularities of human behavior to help other people.

In the second stage, which might be called the period of illu-sion of nonignorance, came the disturbing discovery that, al-though on the one hand I knew that I didn't know what the secret was, suddenly I discovered that on the other hand people were looking to me as though they thought I might know the secret or be closer to the secret than they. . . . None of the research that I did worked nor did any of my activities provide any secret, but again I could always say, "Well, we didn't have enough cases," or "we must improve the methodology," and there were many other state-ments which I am sure you are familiar with. One can postpone the moment of painful discovery but eventually the unhappy truth finally becomes apparent—that although many people may be looking to you and listening to you—you have patients and stu-dents and you're going to PTA meetings and they are looking to you for the secret—still eventually you begin to think maybe, maybe you don't know what you're talking about.

After this rare and revealing admission of doubts that few psy-chotherapists dare state but many have felt, Leary continued at length in describing the various types of research in testing and cat-aloguing and systematizing which had occupied him and his staff. But in this endeavor he was confronted with the problems of no common language and no common unit for observation:

Which natural events are we going to get in permanent form that we can then count? Rather than studying natural free be-havior, I have been experimenting with the possibility of developing standardized languages for the analysis of any natural transaction. Of all the poetic notions and musical notes and lyric strains that we use, words like "progress," "help," and "improvement" are the most far out. We operate with too little information about ourselves and about the other guy. I don't have any theory about new variables in psychology, no new words or language of psychology. I am simply trying to develop new ways of feeding back to human beings what they are doing and the noises they are making. The most exciting thing in the world to me right now is to get at the discrepancies between people involved in the same interaction. Because once you've got that you have a question, "How come?"

He deplored the absence of standardized language in human behavior, noting that stockbrokers, automobile salesmen, and baseball players do better:

Even automobile salesmen have their little blue books and they've really done much better in behavioral science than we people who claim to be the experts. In sports, every baseball player, his natural behavior, is recorded in the form of indices, like his Runs-Batted-In or his Earned-Run-Average. To understand and to make predictions about baseball, if you decide you're going to sell your first baseman to get a right-handed pitcher, you have a raft of behavioral indices. They don't use poetic language like, "He runs after a flyball like a deer," or "He's an obsessive fielder." They tend to use behaviors.

I had been pursuing a myth in trying to find the secret. I wanted to grow up and be a clever therapist and a clever diagnostician. All these hopes of mine were based on the assumption that there are laws, there are regularities, there are secrets, there are techniques which can be applied, and that study and research can bring these secrets to us.

Transactional Analysts claim to have found some of these regularities. We claim to have found a new language of psychology, which

Leary felt such a need of, and we claim to be a great deal closer to the secret of human behavior than we have ever been before.

In this chapter I have presented some of the basic information that has proved useful to a great many people who have been treated in my groups, using Transactional Analysis as an intellectual tool to understand the basis of behavior and feelings. A tool often works better and has more meaning if we have some idea how it was developed, how it is different. Is it derived from authentic data or is it just another theory? Was Berne's book *Games People Play* a best seller because of a fad, or does it offer people some easily understood and authentic ideas about themselves as they reveal their past in the present games they play? In the next chapter we begin the description of this tool, by the definitions of Parent, Adult, and Child. Because these three words have specific and comprehensive meanings different from their usual meanings, Parent, Adult, and Child will be capitalized throughout the book. As you will discover in the next chapter, Parent is not the same as mother or father, Adult means something quite different from a grownup, and Child is not the same as a little person.

2

Parent, Adult, and Child

The passion for truth is silenced by answers which have the
weight of undisputed authority.

—Paul Tillich

Early in his work in the development of Transactional Analysis,
Berne observed that as you watch and listen to people you can see
them change before your eyes. It is a total kind of change. There are
simultaneous changes in facial expression, vocabulary, gestures,
posture, and body functions, which may cause the face to flush, the
heart to pound, or the breathing to become rapid.

We can observe these abrupt changes in everyone: the little
boy who bursts into tears when he can't make a toy work, the teen-
age girl whose woeful face floods with excitement when the phone
finally rings, the man who grows pale and trembles when he gets
the news of a business failure, the father whose face "turns to
stone" when his son disagrees with him. The individual who
changes in these ways is still the same person in terms of bone
structure, skin, and clothes. So what changes inside him? He
changes *from* what *to* what?

This was the question which fascinated Berne in the early development of Transactional Analysis. A thirty-five-year-old lawyer, whom he was treating, said, "I'm not really a lawyer, I'm just a little boy." Away from the psychiatrist's office he was, in fact, a successful lawyer, but in treatment he felt and acted like a little boy. Sometimes during the hour he would ask, "Are you talking to the lawyer or to the little boy?" Both Berne and his patient became intrigued at the existence and appearance of these two real people, or states of being, and began talking about them as "the adult" and "the child." Treatment centered around separating the two. Later another state began to become apparent as a state distinct from "adult" and "child." This was "the parent" and was identified by behavior which was a reproduction of what the patient saw and heard his parents do when he was a little boy.

Changes from one state to another are apparent in manner, appearance, words, and gestures. A thirty-four-year-old woman came to me for help with a problem of sleeplessness, constant worry over "what I am doing to my children," and increasing nervousness. In the course of the first hour she suddenly began to weep and said, "You make me feel like I'm three years old." Her voice and manner were that of a small child. I asked her, "What happened to make you feel like a child?" "I don't know," she responded, and then added, "I suddenly felt like a failure." I said, "Well, let's talk about children, about the family. Maybe we can discover something inside of you that produces these feelings of failure and despair." At another point in the hour her voice and manner again changed suddenly. She became critical and dogmatic: "After all, parents have rights, too. Children need to be shown their place." During one hour this mother changed to three different and distinct personalities: one of a small child dominated by feelings, one of a self-righteous parent, and one of a reasoning, logical, grown-up woman and mother of three children.

Continual observation has supported the assumption that these three states exist in all people. It is as if in each person there is the same little person he was when he was three years old. There are also within him his own parents. These are recordings in the brain of actual experiences of internal and external events, the most significant of which happened during the first five years of life. There is a third state, different from these two. The first two are called Parent and Child, and the third, Adult. (See Figure 1.)

These states of being are not roles but psychological realities. Berne says that "Parent, Adult, and Child are not concepts like Superego, Ego, and Id . . . but phenomenological realities." * The state is produced by the playback of recorded data of events in the past, involving real people, real times, real places, real decisions, and real feelings.

The Parent

The Parent is a huge collection of recordings in the brain of unquestioned or imposed external events perceived by a person in his early years, a period which we have designated roughly as the first five years of life. This is the period before the social birth of the individual, before he leaves home in response to the demands of society and enters school. (See Figure 2.) The name Parent is most descriptive of this data inasmuch as the most significant "tapes" are those provided by the example and pronouncements of his own real parents or parent substitutes. Everything the child saw his parents do and everything he heard them say is recorded in the Parent. Everyone has a Parent in that everyone experienced external stimuli in the first five years of life. Parent is specific for every person, being the recording of that set of early experiences unique to him.

The data in the Parent was taken in and recorded "straight" without editing. The situation of the little child, his dependency,

*E. Berne, *Transactional Analysis in Psychotherapy* (New York: Grove Press, 1961), p. 24.

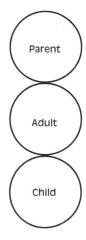

Fig. 1. The Personality.

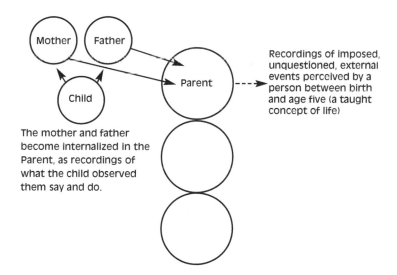

Recordings of imposed, unquestioned, external events perceived by a person between birth and age five (a taught concept of life)

The mother and father become internalized in the Parent, as recordings of what the child observed them say and do.

Fig. 2. The Parent.

and his inability to construct meanings with words made it impossible for him to modify, correct, or explain. Therefore, if the parents were hostile and constantly battling each other, a fight was recorded with the terror produced by seeing the two persons on whom the child depended for survival about to destroy each other. There was no way of including in this recording the fact that the father was inebriated because his business had just gone down the drain or that the mother was at her wits' end because she had just found she was pregnant again.

In the Parent are recorded all the admonitions and rules and laws that the child heard from his parents and saw in their living. They range all the way from the earliest parental communications, interpreted nonverbally through tone of voice, facial expression, cuddling, or non-cuddling, to the more elaborate verbal rules and regulations espoused by the parents as the little person became able to understand words. In this set of recordings are the thousands of "no's" directed at the toddler, the repeated "don'ts" that bombarded him, the looks of pain and horror in mother's face when his clumsiness brought shame on the family in the form of Aunt Ethel's broken antique vase.

Likewise are recorded the coos of pleasure of a happy mother and the looks of delight of a proud father. When we consider that the recorder is on all the time we begin to comprehend the immense amount of data in the Parent. Later come the more complicated pronouncements: Remember, Son, wherever you go in the world you will always find the best people are Methodists; never tell a lie; pay your bills; you are judged by the company you keep; you are a good boy if you clean your plate; waste is the original sin; you can never trust a man; you can never trust a woman; you're damned if you do and damned if you don't; you can never trust a cop; busy hands are happy hands; don't walk under ladders; do unto others as you would have them do unto you; do others in that they don't do you in.

The significant point is that whether these rules are good or bad in the light of a reasonable ethic, they are recorded as *truth* from the source of all security, the people who are "six feet tall" at a time when it is important to the two-foot-tall child that he please and obey them. It is a permanent recording. A person cannot erase it. It is available for replay throughout life.

This replay is a powerful influence throughout life. These examples—coercing, forcing, sometimes permissive but more often restrictive—are rigidly internalized as a voluminous set of data essential to the individual's survival in the setting of a group, beginning with the family and extending throughout life in a succession of groups necessary to life. Without a physical Parent the child would die. The internal Parent also is lifesaving, guarding against many dangers which, perceived experientially, could cause death. In the Parent is the recording, "Don't touch that knife!" It is a thunderous directive. The threat to the little person, as he sees it, is that his mother will spank him or otherwise show disapproval. The greater threat is that he can cut himself and bleed to death. He cannot perceive this. He does not have adequate data. The recording of parental dictates, then, is an indispensable aid to survival, in both the physical and the social sense.

Another characteristic of the Parent is the fidelity of the recordings of inconsistency. Parents say one thing and do another. Parents say, "Don't lie," but tell lies. They tell children that smoking is bad for their health but smoke themselves. They proclaim adherence to a religious ethic but do not live by it. It is not safe for the little child to question this inconsistency, and so he is confused. Because this data causes confusion and fear, he defends himself by turning off the recording.

We think of the Parent predominantly as the recordings of the transactions between the child's two parents and their transactions with him. It may be helpful to consider the recordings of Parent data as somewhat like the recording of stereophonic sound. There are two sound tracks that, if harmonious, produce a beautiful effect when played together. If they are not harmonious, the effect is unpleasant and the recording is put aside and played very little, if at all. This is what happens when the Parent contains discordant material. The Parent is repressed or, in the extreme, blocked out altogether. Mother may have been a "good" mother and father may have been "bad," or vice versa. There is much useful data which is stored as a result of the transmission of good material from one parent; but since the Parent does contain material from the other parent that is contradictory and productive of anxiety, the Parent as a whole is weakened or fragmented. Parent data that is discordant is not allowed to come on "audibly" as a strong influence in the person's life.

Another way to describe this phenomenon is to compare it with the algebraic equation: a plus times a minus equals a minus. It does not matter how big the plus was, or how little the minus was. The result is always a minus—a weakened, disintegrated Parent. The effect in later life may be ambivalence, discord, and despair— for the person, that is, who is not free to examine the Parent.

Much Parent data appears in current living in the "how-to" category: how to hit a nail, how to make a bed, how to eat soup, how to blow your nose, how to thank the hostess, how to shake hands, how to pretend no one's at home, how to fold the bath towels, or how to dress the Christmas tree. The *how to* comprises a vast body of data acquired by watching the parents. It is largely useful data which makes it possible for the little person to learn to get along by himself. Later (as his Adult becomes more skillful and free to examine Parent data) these early ways of doing things may be updated and replaced by better ways that are more suited to a changed reality. A person whose early instructions were accompanied by stern intensity may find it more difficult to examine the old ways and may hang onto them long after they are useful, having developed a compulsion to do it "this way and no other."

The mother of a teen-ager related the following parental edict, which had long governed her housekeeping procedures. Her mother had told her, "You *never* put a hat on a table or a coat on a bed." So she went through life never putting a hat on a table or a coat on a bed. Should she occasionally forget, or should one of her youngsters break this old rule, there was an over-reaction that ✓ seemed inappropriate to the mere violation of the rules of simple neatness. Finally, after several decades of living with this unexamined law, mother asked grandmother (by then in her eighties), "Mother, *why* do you never put a hat on a table or a coat on a bed?"

Grandmother replied that when she was little there had been some neighbor children who were "infested," and her mother had warned her that it was important they never put the neighbor children's hats on the table or their coats on the bed. Reasonable enough. The urgency of the early admonition was understandable. In terms of Penfield's findings it was also understandable why the recording came on with the original urgency. Many of the rules we live by are like this.

Some influences are more subtle. One modern housewife with every up-to-date convenience in her home found she simply did not have any interest in buying a garbage-disposal unit. Her husband encouraged her to get one, pointing out all the reasons this would simplify her kitchen procedures. She recognized this but found one excuse after another to postpone going to the appliance store to select one. Her husband finally confronted her with his belief that she was *deliberately* not getting a garbage disposal. He insisted she tell him why.

A bit of reflection caused her to recognize an early impression she had about garbage. Her childhood years were the Depression years of the 1930s. In her home, garbage was carefully saved and fed to the pig, which was butchered at Christmas and provided an important source of food. The dishes were even washed without soap so that the dishwater, with its meager offering of nutrients, could be included in the slops. As a little girl she perceived that garbage was important, and as a grown woman she found it difficult to rush headlong into purchasing a new-fangled gadget to dispose of it. (She bought the disposal unit and lived happily ever after.)

When we realize that thousands of these simple rules of living are recorded in the brain of every person, we begin to appreciate what a comprehensive, vast store of data the Parent includes. Many of these edicts are fortified with such additional imperatives as "never" and "always" and "never forget that" and, we may assume, pre-empt certain primary neurone pathways that supply ready data for today's transactions. These rules are the origins of compulsions and quirks and eccentricities that appear in later behavior. Whether Parent data is a burden or a boon depends on how appropriate it is to the present, on whether or not it has been updated by the Adult, the function of which we shall discuss in this chapter.

There are sources of Parent data other than the physical parents. A three-year-old who sits before a television set many hours a day is recording what he sees. The programs he watches are a "taught" concept of life. If he watches programs of violence, I believe he records violence in his Parent. That's how it is. That is life! This conclusion is certain if his parents do not express opposition by switching the channel. If they enjoy violent programs the youngster gets a double sanction—the set and the folks—and he assumes permission to be violent provided he collects the required amount of injustices. The little person collects his own reasons to shoot up the place, just as the sheriff does; three nights of cattle rustlers, a stage holdup, and a stranger foolin' with Miss Kitty can be easily matched in the life of the little person. Much of what is experienced at the hands of older siblings or other authority figures also is recorded in the Parent. Any external situation in which the little person feels himself to be dependent to the extent that he is not free to question or to explore produces data which is stored in the Parent. (There is another type of external experience of the very small child which is not recorded in the Parent, and which we shall examine when we describe the Adult.)

The Child

While external events are being recorded as that body of data we call the Parent, there is another recording being made simultaneously. This is the recording of *internal* events, the responses of the little person to what he sees and hears. (Figure 3.) In this connection it is important to recall Penfield's observation that the subject feels again the emotion which the situation originally produced in him, and he is aware of the same interpretations, true or false, which he himself gave to the experience in the first place. Thus, evoked recollection is not the exact photographic or phonographic reproduction of past scenes or events. It is reproduction of what the patient *saw and heard and felt and understood.** [Italics mine.]

It is this "seeing and hearing and feeling and understanding" body of data which we define as the Child. Since the little person has no vocabulary during the most critical of his early experiences,

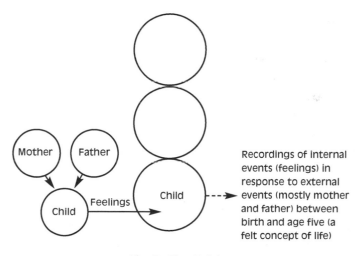

Recordings of internal events (feelings) in response to external events (mostly mother and father) between birth and age five (a felt concept of life)

Fig. 3. The Child.

*W. Penfield, "Memory Mechanisms," *A.M.A. Archives of Neurology and Psychiatry* 67 (1952): 178–198, with discussion by L. S. Kubie et al.

most of his reactions are *feelings.* We must keep in mind his situation in these early years. He is small, he is dependent, he is inept, he is clumsy, he has no words with which to construct meanings. Emerson said we "must know how to estimate a sour look." The child does not know how to do this. A sour look turned in his direction can only produce feelings that add to his reservoir of negative data about himself. *It's my fault. Again. Always is. Ever will be. World without end.*

During this time of helplessness there are an infinite number of total and uncompromising demands on the child. On the one hand, he has the urges (genetic recordings) to empty his bowels ad lib., to explore, to know, to crush and to bang, to express feelings, and to experience all of the pleasant sensations associated with movement and discovery. On the other hand, there is the constant demand from the environment, essentially the parents, that he give up these basic satisfactions for the reward of parental approval. This approval, which can disappear as fast as it appears, is an unfathomable mystery to the child, who has not yet made any certain connection between cause and effect.

The predominant by-product of the frustrating, civilizing process is negative feelings. On the basis of these feelings the little person early concludes, "I'm not OK." We call this comprehensive self-estimate the NOT OK, or the NOT OK Child. This conclusion and the continual experiencing of the unhappy feelings which led to it and confirm it are recorded permanently in the brain and cannot be erased. This permanent recording is the residue of having been a child. Any child. Even the child of kind, loving, well-meaning parents. It is the *situation of childhood* and *not* the intention of the parents which produces the problem. (This will be discussed at length in the next chapter, about life positions.) An example of the dilemma of childhood was a statement made by my seven-year-old daughter, Heidi, who one morning at breakfast said, "Daddy, when I have an OK Daddy and an OK Mama, how come *I'm* not OK?"

When the children of "good" parents carry the NOT OK burden, one can begin to appreciate the load carried by children whose parents are guilty of gross neglect, abuse, and cruelty. As in the case of the Parent, the Child is a state into which a person may be transferred at almost any time in his current transactions. There are many things that can happen to us today which recreate the situation of childhood and bring on the same feelings we felt then. Frequently we may find ourselves in situations where we are faced with impossible alternatives, where we find ourselves in a corner, either actually, or in the way we see it. These "hook the Child," as we say, and cause a replay of the original feelings of frustration, rejection, or abandonment, and we relive a latter-day version of the small child's primary depression. Therefore, when a person is in the grip of feelings, we say his Child has taken over. When his anger dominates his reason, we say his Child is in command.

There is a bright side, too! In the Child is also a vast store of positive data. In the Child reside creativity, curiosity, the desire to explore and know, the urges to touch and feel and experience, and the recordings of the glorious, pristine feelings of first discoveries. In the Child are recorded the countless, grand *a-ha* experiences, the firsts in the life of the small person, the first drinking from the garden hose, the first stroking of the soft kitten, the first sure hold on mother's nipple, the first time the lights go on in response to his flicking the switch, the first submarine chase of the bar of soap, the repetitious going back to do these glorious things again and again. The feelings of these delights are recorded, too. With all the NOT OK recordings, there is a counterpoint, the rhythmic OK of mother's rocking, the sentient softness of the favorite blanket, a continuing good response to favorable external events (if this is indeed a favored child), which also is available for replay in today's transactions. This is the flip side, the happy child, the carefree, butterfly-chasing little boy, the little girl with chocolate on her face. This comes on in today's transactions, too. However, our observations both of small children and of ourselves as grownups convince us that the NOT OK feelings generally outweigh the good. This is why we believe it is a fair estimate to say that everyone has a NOT OK Child.

Frequently I am asked, When do the Parent and Child stop recording? Do the Parent and Child contain only experiences in the first five years of life? I believe that by the time the child leaves the home for his first independent social experience—school—he has been exposed to nearly every possible attitude and admonition of his parents, and thenceforth further parental communications are essentially a reinforcement of what has already been recorded. The fact that he now begins to "use his Parent" on others also has a reinforcing quality in line with the Aristotelian idea that that which is expressed is impressed. As to further recordings in the Child, it is hard to imagine that any emotion exists which has not already been felt in its most intense form by the time the youngster is five years old. This is consistent with most psychoanalytic theory, and, in my own observation, is true.

If, then, we emerge from childhood with a set of experiences which are recorded in an inerasable Parent and Child, what is our hope for change? How can we get off the hook of the past?

The Adult

At about ten months of age a remarkable thing begins to happen to the child. Until that time his life has consisted mainly of helpless or unthinking responses to the demands and stimulations by those around him. He has a Parent and a Child. What he has not had is the ability either to choose his responses or to manipulate his surroundings. He has had no self-direction, no ability to move out to meet life. He has simply taken what has come his way.

At ten months, however, he begins to experience the power of locomotion. He can manipulate objects and begins to move out, freeing himself from the prison of immobility. It is true that earlier, as at eight months, the infant may frequently cry and need help in getting out of some awkward position, but he is unable to get out of it by himself. At ten months he concentrates on inspection and exploitation of toys. According to the studies conducted by Gesell and Ilg, the ten-month-old child

> . . . enjoys playing with a cup and pretends to drink. He brings objects to his mouth and chews them. He enjoys gross motor activity: sitting and playing after he has been set up, leaning far forward, and re-erecting himself. He secures a toy, kicks, goes from sitting to creeping, pulls himself up, and may lower himself. He is beginning to cruise. Social activities which he enjoys are peek-a-boo and lip play, walking with both hands held, being put prone on the floor, or being placed in a rocking toy. Girls show their first signs of coyness by putting their heads to one side as they smile.*

The ten-month-old has found he is able to do something which grows from his own awareness and original thought. This self-actualization is the beginning of the Adult. (Figure 4.) Adult data accumulates as a result of the child's ability to find out for himself what is different about life from the "taught concept" of life in his Parent and the "felt concept" of life in his Child. The Adult develops a "thought concept" of life based on data gathering and data processing.

The motility which gives birth to the Adult becomes reassuring in later life when a person is in distress. He goes for a walk to "clear his mind." Pacing is seen similarly as a relief from anxiety. There is a recording that movement is good, that it has a separating quality, that it helps him see more clearly what his problem is.

The Adult, during these early years, is fragile and tentative. It is easily "knocked out" by commands from the Parent and fear in

*Arnold Gesell and Francis L. Ilg, *Infant and Child in the Culture of Today* (New York: Harper, 1943), pp. 116-122.

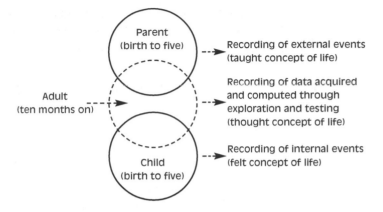

Fig. 4.   Gradual emergence of the Adult
beginning at ten months.

the Child. Mother says about the crystal goblet, "No, no! Don't touch that!" The child may pull back and cry, but at the first opportunity he will touch it anyway to see what it is all about. In most persons the Adult, despite all the obstacles thrown in its way, survives and continues to function more and more effectively as the maturation process goes on.

The Adult is "principally concerned with transforming stimuli into pieces of information, and processing and filing that information on the basis of previous experience.* It is different from the Parent, which is "judgmental in an imitative way and seeks to enforce sets of borrowed standards, and from the Child, which tends to react more abruptly on the basis of prelogical thinking and poorly differentiated or distorted perceptions." Through the Adult the little person can begin to tell the difference between life as it was taught and demonstrated to him (Parent), life as he felt it or wished it or fantasied it (Child), and life as he figures it out by himself (Adult).

The Adult is a data-processing computer, which grinds out decisions after computing the information from three sources: the

*Berne, *Transactional Analysis in Psychotherapy.*

Parent, the Child, and the data which the Adult has gathered and is gathering (Figure 5). One of the important functions of the Adult is to examine the data in the Parent, to see whether or not it is true and still applicable today, and then to accept it or reject it; and to examine the Child to see whether or not the feelings there are appropriate to the present or are archaic and in response to archaic Parent data. The goal is not to do away with the Parent and Child but to be free to examine these bodies of data. The Adult, in the words of Emerson, "must not be hindered by the name of goodness, but must examine if it be goodness"; or badness, for that matter, as in the early decision, "I'm not OK."

The Adult testing of Parent data may begin at an early age. A secure youngster is one who finds that most Parent data is reliable: "They told me the truth!"

It really *is* true that cars in the street are dangerous," concludes the little boy who has seen his pet dog hurt by a car in the street. It

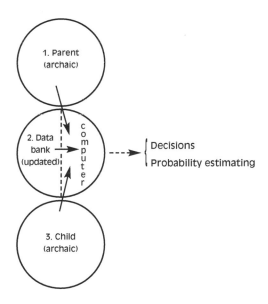

Fig. 5. The Adult gets data from three sources.

really *is* true that things go better when I share my toys with Bobby," thinks the little boy who has been given a prized possession by Bobby. "It really *does* feel better when my pants aren't wet," concludes the little girl who has learned to go to the bathroom by herself. If parental directives are grounded in reality, the child, through his own Adult, will come to realize integrity, or a sense of wholeness. What he tests holds up under testing. The data which he collects in his experimentation and examination begins to constitute some "constants" that he can trust. His findings are supported by what he was taught in the first place.

It is important to emphasize that the verification of Parent data does not erase the NOT OK recordings in the Child, which were produced by the early imposition of this data. Mother believes that the only way to keep three-year-old Johnny out of the street is to spank him. He does not understand the danger. His response is fear, anger, and frustration with no appreciation of the fact that his mother loves him and is protecting his life. The fear, anger, and frustration are recorded. These feelings are not erased by the later understanding that she was right to do what she did, but the understanding of how the original situation of childhood produced so many NOT OK recordings of this type can free us of their continual replay in the present. *We cannot erase the recording, but we can* *choose to turn it off!*

In the same way that the Adult updates Parent data to determine what is valid and what is not, it updates Child data to determine which feelings may be expressed safely. In our society it is considered appropriate for a woman to cry at a wedding, but it is not considered appropriate for that woman to scream at her husband afterward at the reception. Yet both crying and screaming are emotions in the Child. The Adult keeps emotional expression appropriate. The Adult's function in updating the Parent and Child is diagramed in Figure 6. The Adult within the Adult in this figure refers to updated reality data. (The evidence once told me space travel was only fantasy; now I know it is reality.)

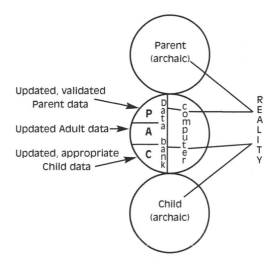

Fig. 6. The updating function of the Adult through reality testing.

Another of the Adult's functions is *probability estimating*. This function is slow in developing in the small child and, apparently, for most of us, has a hard time catching up throughout life. The little person is constantly confronted with unpleasant alternatives (either you eat your spinach or you go without ice cream), offering little incentive for examining probabilities. Unexamined probabilities can underlie many of our transactional failures, and unexpected danger signals can cause more Adult "decay," or delay, than expected ones. There are similarities here to the stock ticker in investment concerns, which may run many hours behind on very active trading days. We sometimes refer to this delay as "computer lag," a remedy for which is the old, familiar practice of "counting to ten."

The capacity for probability estimating can be increased by conscious effort. Like a muscle in the body, the Adult grows and increases in efficiency through training and use. If the Adult is alert to the possibility of trouble, through probability estimating, it can also devise solutions to meet the trouble if and when it comes.

Under sufficient stress, however, the Adult can be impaired to the point where emotions take over inappropriately. The boundaries between Parent, Adult, and Child are fragile, sometimes indistinct, and vulnerable to those incoming signals which tend to recreate situations we experienced in the helpless, dependent days of childhood. The Adult sometimes is flooded by signals of the "bad news" variety so overwhelming that the Adult is reduced to an "onlooker" in the transaction. An individual in this situation might say, "I knew what I was doing was wrong, but I couldn't help myself."

Unrealistic, irrational, non-Adult responses are seen in a condition referred to as traumatic neurosis. The danger, or "bad news" signal, hits the Parent and the Child at the same time it hits the Adult. The Child responds in the way it originally did, with a feeling of NOT OK. This may produce all kinds of regressive phenomena. The individual may again feel himself to be a tiny, helpless, dependent child. One of the most primitive of these phenomena is thought blocking. One place this can be seen is in psychiatric hospitals that have a locked-door policy. When the door is locked on a new patient, his retreat is rapid and pronounced. This is why I am opposed to treating patients in a setting where the emphasis is on parental care. Catering to the helpless Child in the individual delays the reconstructive process of restoring the Adult to the executive function.

An ideal hospital would be a comfortable motel with "play area" for the Child, surrounding a clinic building devoted to activities designed for achieving autonomy of the Adult. The nurses would not wear uniforms or serve as parents to the patients. Instead, nurses in street clothing would apply their skills and training to help each individual learn the identity of his Parent, Adult, and Child.

In our treatment groups we use certain colloquial catch phrases such as, "Why don't you stay in your Adult?" when a member finds his feelings are taking over. Another of these is,

"What was the original transaction?" This is asked as a means of "turning on the Adult" to analyze the similarity between the present incoming signal producing the present distress and the original transaction, in which the small child experienced distress.

The ongoing work of the Adult consists, then, of checking out old data, validating or invalidating it, and refiling it for future use. If this business goes on smoothly and there is a relative absence of conflict between what has been taught and what is real, the computer is free for important new business, *creativity*. Creativity is born from curiosity in the Child, as is the Adult. The Child provides the "want to" and the Adult provides the "how to." The essential requirement for creativity is computer time. If the computer is cluttered with old business there is little time for new business. Once checked out, many Parent directives become automatic and thus free the computer for creativity. Many of our decisions in day-to-day transactions are automatic. For instance, when we see an arrow pointing down a one-way street, we automatically refrain from going the opposite way. We do not involve our computer in lengthy data processing about highway engineering, the traffic death toll, or how signs are painted. Were we to start from scratch in every decision or operate entirely without the data that was supplied by our parents, our computer would rarely have time for the creative process.

Some people contend that the undisciplined child, unhampered by limits, is more creative than the child whose parents set limits. I do not believe this is true. A youngster has more time to be creative—to explore, invent, take apart, and put together—if he is not wasting time in futile decision making for which he has inadequate data. A little boy has more time to build a snowman if he is not allowed to engage Mother in a long hassle about whether or not to wear overshoes. If a child is allowed to be creative by painting the front room walls with shoe polish, he is unprepared for the painful consequences when he does so at the neighbor's house. Painful outcomes do not produce OK feelings. There are other consequences that take time, such as mending in the hospital after a trial-and-error encounter with a car in the street. There is just so much computer time. Conflict uses a great deal. An extremely time-consuming conflict is produced when what parents say is true does not seem to be true to the Adult. The most creative individual is the one who discovers that a large part of the content of the Parent squares with reality. He can then file away this validated information in the Adult, trust it, forget about it, and get on with other things—like how to make a kite fly, how to build a sand castle, or how to do differential calculus.

However, many youngsters are preoccupied much of the time with the conflict between Parent data and what they see as reality. Their most troubling problem is that they do not understand why the Parent has such a hold on them. When Truth comes to knock at the Parent's door, the Parent says, "Come, let us reason together." The little child whose father is in jail and whose mother steals to support him may have a loud recording in his Parent, "You never trust a cop!" So he meets a friendly one. His Adult computes all the data about this nice guy, how he gets the ball game started in the sand lot, how he treats the gang to popcorn, how he is friendly, and how he speaks in a quiet voice. For this youngster there is conflict. What he sees as reality is different from what he has been taught. The Parent tells him one thing and the Adult another. During the period of his actual dependency upon his parents for security, however tenuous this security may be, it is likely he will accept the parents' verdict that cops are bad. This is how prejudice is transmitted. *For a little child, it may be safer to believe a lie than to believe his own eyes and ears.* The Parent so threatens the Child (in a continuing internal dialogue) that the Adult gives up and stops trying to inquire into areas of conflict. Therefore, "cops are bad" comes through as truth. This is called *contamination* of the Adult and will be examined in Chapter 6.

8
8
3

The Four Life Positions

For the sadness in legitimate humor consists in the fact that
honestly and without deceit it reflects in a purely human way
upon what it is to be a child.

—**Søren Kierkegaard**

Very early in life every child concludes, "I'm not OK." He
makes a conclusion about his parents, also: "You're OK." This is
the first thing he figures out in his life-long attempt to make sense
of himself and the world in which he lives. This position, I'M NOT
OK—YOU'RE OK, is the most deterministic decision of his life. It is
permanently recorded and will influence everything he does. Because it is a decision it can be changed by a new decision. But not
until it is understood.

In order to support these contentions I wish to devote the
first part of this chapter to an examination of the situations of the
newborn, the young infant, and the growing child, in both the
preverbal and verbal years. Many people insist they had a "happy
childhood" and concluded nothing like I'M NOT OK—YOU'RE OK.
I believe strongly that *every* child concludes it, "happy childhood" notwithstanding. First, I wish to examine the situation of

his entry into life and to point to the evidence that the events of his birth and his infant life are recorded, even though they are not remembered.

In this connection we note again Penfield's conclusions that the brain performs three functions: (1) recording, (2) recalling, and (3) reliving. Although *recall* from the earliest period of life is not possible, we have evidence that we can and do *relive* the earliest experiences in the form of returning to the feeling state of the newborn infant. Because the infant cannot use words, his reactions are limited to sensations, feelings, and perhaps vague, archaic fantasies. His feelings are expressed by crying or by various body movements that indicate either distress or comfort. His sensations and fantasies, though ineffable because of his wordlessness at the time they were recorded, do replay occasionally in dreams in later life.

To illustrate: A patient reported a dream that had recurred throughout her life. Each time she had this dream, she awoke in a state of extreme panic, with rapid heartbeat and heavy breathing. She struggled to describe the dream but she could not find words for it. In one attempt at description she said she thought she felt as if she were "just a tiny, little, small speck, and big, huge, round, cosmic things were swirling around me, like great spirals, getting bigger and bigger, and threatening to engulf me, and I just seemed to disappear in this vast, enormous thing." Though her report was accompanied by her observation about losing her identity, the nature of the extreme panic would seem to indicate there may have been a fear of losing her life, as a primary biological reaction to the threat of death.

Some time later she again reported the dream. It was the first time she had dreamed it for about a year. She had been traveling, and she and her husband had eaten lunch in an out-of-the-way restaurant with an atmosphere of a higher quality than the food's. She did not feel well when they returned to their hotel, so she lay down for a nap. She fell asleep. It was not long before she awoke in the panic of this same dream. She also had severe stomach cramps, which "had me all doubled up in pain." No recent event had been particularly anxiety-provoking, and the panic dream seemed to have some direct connection with the extreme, primordial gut pain. The dream was still indescribable; however, she did report another sensation, the feeling that she was suffocating.

Certain information about the patient's mother helped to suggest a possible origin of this dream. The mother, a large, plump woman, had breast-fed her children and had held to the idea there was no problem that eating would not cure. Her idea of well-cared-for children was well-fed children. She also was an aggressive, domineering woman. We deduced (which is all we can do) that the dream had its origin in a time before the patient had words, since she could not describe the content. The association with the belly cramp suggested some connection with an early eating experience. The probability is that if, as an infant, the patient had had enough, or had had a full feeling and quit nursing, the mother would insist she have more. (This was before the era of demand-feeding: "Fill up now, it will have to last you.") Feelings of "dream-state" sleepiness, suffocation, and stomach cramps could have been present. The content (the small thing being engulfed by huge, cosmic things) could have been a replay of the infant's perception of her situation—herself, the small speck, being engulfed by the huge, round things, mother's breasts, or the huge presence of the mother herself.

This type of dream material lends support to the assumption that *our earliest experiences, though ineffable, are recorded and do replay in the present.* Another indication that experiences are recorded from the time of birth is the retention of past gains. The infant's responses to external stimuli, although at first instinctual, soon reflect conditioned or learned (or recorded) experience. For instance, he learns to look in the direction of mother's footsteps. If all experiences and feelings are recorded, we can understand the extreme panic, or rage, or fear we feel in certain situations today as a reliving of the original state of panic or rage or fear that we felt as infants. We can think of this as a replay of the original tape.

To understand the implications, it is important to examine the situation of the infant. In reference to Figure 7, we see a line representing a span from the moment of conception to the age of five. The first block of time is the nine months between conception and biological birth. During these nine months there occurred a beginning of life in the most perfect environment the human individual may ever experience. This way of life is referred to as a state of symbiotic intimacy.

Then, at biological birth, the little individual, within the brief span of a few hours, is pushed out into a state of catastrophic contrast in which he is exposed to foreign and doubtless terrifying extremes of cold, roughness, pressure, noise, nonsupport, brightness, separateness, and abandonment. The infant is, for a short time, cut off, apart, separate, unrelated. Common to the many theories about the birth trauma is the assumption that the feelings produced by this event were recorded and reside in some form in the brain. This assumption is supported by the great number of repetitious dreams of the "drainage pipe" variety which so many individuals experience following situations of extreme stress. The patient describes a dream in which he is swept from a body of water of relative calm into a sewer or drainage pipe. He experiences the feeling of increasing velocity and compression. This feeling also is experienced in the state of claustrophobia. The infant is flooded with overwhelming, unpleasant stimulations, and the feelings resulting in the child are, according to Freud, the model for all later anxiety.*

Within moments the infant is introduced to a rescuer, another human being who picks him up, wraps him in warm coverings, supports him, and begins the comforting act of "stroking." This is the point (Figure 7) of Psychological Birth. This is the first incom-

*Sigmund Freud, *The Problem of Anxiety* (New York: Norton, 1936).

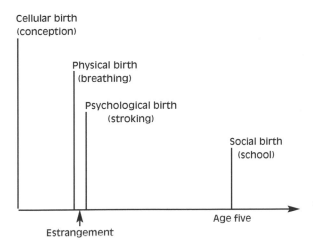

Fig. 7. Births of the individual from conception to age five.

ing data that life "out there" isn't all bad. It is a reconciliation, a reinstatement of closeness. It turns on his will to live. Stroking, or repetitious bodily contact, is essential to his survival. Without it he will die, if not physically, then psychologically. Physical death from a condition known as marasmus once was a frequent occurrence in foundling homes where there was a deprivation of this early stroking. There was no physical cause to explain these deaths except the absence of essential stimulation.

This painful on-again-off-again keeps the infant in a constant state of disequilibrium. During the first two years of life he does not have conceptual "thinking" tools—*words*—to construct an explanation of his uncertain status in his world. He is, however, continually recording the feelings which grow from the relationship between himself and others, primarily mother, and these feelings are directly related to stroking and nonstroking. Whoever provides stroking is OK. His estimate of himself is unsure because his OK feelings are transitory and continually being replaced by NOT OK feelings. Finally the uncertainty convinces him I'M NOT OK. At what

point does the child make final his decision as to the position, I'M
NOT OK—YOU'RE OK?

Piaget,* on the basis of meticulous observations of infants
and small children, believes that the development of causality
(what follows what) begins in the early months of life and is acquired by the end of the second year. In other words, data, in the
form of a jumble of impressions, begin accumulating in certain
sequential patterns, to a point where a preverbal position, or conclusion, is possible. Piaget says: "In the course of the first two
years of childhood the evolution of sensorimotor intelligence,
and also the correlative elaboration of the universe, seem to lead
to a *state of equilibrium* bordering on rational thought." I believe
this state of equilibrium, evident at the end of the second year or
during the third year, is the product of the child's conclusion
about himself and others: *his life position.* Once his position is
decided he has something solid to work with, some basis for predictability. Piaget says that these early mental processes are not
capable of "knowing or stating truths" but are limited to desiring
success or practical adaptation: If I'M NOT OK and YOU'RE OK,
what can I do to make you, an OK person, be good to me, a NOT
OK person? The position may seem unfavorable, but it is a true
impression, to the child, and it is better than nothing. Thus the
state of equilibrium. The Adult in the little person has achieved
its first mastery in "making sense of life," in solving what Adler
called "life's central problem"—the attitude toward others—and
what Sullivan called the "self-attitudes which are carried forever
by the individual."

One of the clearest statements on the development of positions
is made by Kubie:

> It is possible to make one certain deduction: namely, that
> early in life, sometimes within the earliest months and some-

*Jean Piaget, *The Construction of Reality in the Child* (New York: Basic Books,
1954).

times later, a *central emotional position is frequently established*. . . . The clinical fact which, is already evident is that once a central emotional position is established early in life, it becomes the affective position *to which that individual will tend to return automatically for the rest of his days.* This in turn may constitute either the major safeguard or the major vulnerability of his life. In fact the establishing of a central emotional position may turn out to be one of the *earliest among the universals* in the evolution of the human neurotic process, since it may start even in the pre-verbal and largely pre-symbolic days of infancy. . . . Whenever the central emotional position is painful . . . the individual may spend his whole life defending himself against it, again using conscious, pro-conscious, and unconscious devices whose aim it is to avoid this pain-filled central position.* [Italics mine.]

Kubie then raises the question as to whether or not these positions are alterable later in life. I believe they are. Although the early experiences which culminated in the position cannot be erased, I believe the early positions can be changed. *What was once decided can be undecided.*

Transactional Analysis constructs the following classification of the four possible life positions held with respect to oneself and others:

1. I'M NOT OK—YOU'RE OK
2. I'M NOT OK—YOU'RE NOT OK
3. I'M OK—YOU'RE NOT OK
4. I'M OK—YOU'RE OK

Before I elaborate each position I wish to state a few general observations about positions. I believe that by the end of the second year of life, or sometime during the third year, the child has

*L. S. Kubie, "The Neurotic Process as the Focus of Physiological and Psycho-analytic Research," *The Journal of Mental Science,* Vol. 104, No. 435 (1958).

decided on one of the first three positions. The I'M NOT OK—YOU'RE OK is the first tentative decision based on the experiences of the first year of life. By the end of the second year it is either confirmed and settled or it gives way to Position 2 or 3: I'M NOT OK—YOU'RE NOT OK or I'M OK—YOU'RE NOT OK. Once finalized, the child stays in his chosen position and it governs everything he does. It stays with him the rest of his life, unless he later consciously changes it to the fourth position. People do not shift back and forth. The decision as to the first three positions is based totally on stroking and nonstroking. The first three are nonverbal decisions. They are conclusions, not explanations. Yet they are more than conditioned responses. They are what Piaget calls intellectual elaborations in the construction of causality. In other words, they are a product of Adult data processing in the very little person.

I'm Not OK—You're OK

This is the universal position of early childhood, being the infant's logical conclusion from the situation of birth and infancy. There is OK-ness in this position, because stroking is present. Every child is stroked in the first year of life simply by the fact that he had to be picked up to be cared for. Without at least minimal handling the infant would not survive. There is also NOT-OK-ness. That is the conclusion about himself. I believe the evidence points to the overwhelming accumulation of NOT OK feelings in the child, making logical (on the basis of the evidence *he* has) his NOT OK conclusion about himself. In explaining Transactional Analysis to patients and nonpatients I have found a generally *that's it!* response to the explanation of the origin and existence of the NOT OK Child. I believe that acknowledging the NOT OK Child in each of us is the only sympathetic, thus curative, way games can be analyzed. Considering the universality of games, the universality of the I'M NOT OK is a reasonable deduction. Adler's break with Freud was over this point: sex was not at the basis of man's struggle in life, but rather feelings of inferiority, or NOT OK, which were apparent universally. He claimed that the child, by virtue of his small size and helplessness, inevitably considered himself inferior to the adult figures in his environment. Harry Stack Sullivan was greatly influenced by Adler, and I was greatly influenced by Sullivan, with whom I studied for the five years preceding his death. Sullivan, whose central contribution to psychoanalytic thought was the concept of "interpersonal relationships," or transactions, claimed that the child built his self-estimate totally on the appraisal of others, what he called "reflected appraisals." He said:

> The child lacks the equipment and experience necessary to form an accurate picture of himself, so his only guide is the reactions of others to him. There is very little cause for him to question these appraisals, and in any case he is far too helpless to challenge them or to rebel against them. He passively accepts the judgments, which are communicated empathetically at first, and by words, gestures, and deeds in this period . . . thus the self-attitudes learned early in life are carried forever by the indi- ✓ vidual, with some allowance for the influence of extraordinary environmental circumstances and modification through later experiences.*

In the first position the person feels at the mercy of others. He feels a great need for stroking, or recognition, which is the psychological version of the early physical stroking. In this position there is hope because there is a source of stroking—YOU'RE OK— even if the stroking is not constant. The Adult has something to work on: What must I do to gain their strokes, or their approval? There are two ways in which people may attempt to live out this position.

*From G. S. Blum, *Psychoanalytic Theories of Personality* (New York: McGraw-Hill, 1953), pp. 73, 74.

The first is to live out a *life script** that confirms the NOT OK. It is written unconsciously by the Child. The script may call for a life of withdrawal, since it is too painful to be around OK people. These people may seek stroking through make-believe and engage in an elaborate wish-life of *if I* and *when I.* Another person's script may call for behavior which is provoking to the point where others turn on him (negative stroking), thus proving once again, I'M NOT OK. This is the case of the "bad little boy." *You say I'm bad so I'll be bad!* He may kick and spit and claw his way through life and thus achieve a fraudulent integrity with at least one constant he can count on: I'M NOT OK—YOU'RE OK. There is a kind of miserable sense in this, in that the integrity of the position is maintained, but it leads to despair. The ultimate resolution of this position is giving up (institutionalization) or suicide.

A more common way to live out this position is by a *counter-script* (also unconscious) with borrowed lines from the Parent: YOU CAN BE OK, IF. Such a person seeks friends and associates who have a big Parent because he needs big strokes, and the bigger the Parent, the better the strokes. (OK strokes can only come from OK people, and the Parent is OK, as it was in the beginning.) This person is eager, willing, and compliant to the demands of others. "Some of our best people" are where they are because of these efforts to gain approval. However, they are committed to a lifetime of mountain climbing, and when they reach the top of one mountain they are confronted by still another mountain. The NOT OK writes the script; the YOU'RE OK (and I want to be like you) writes the counter-script. Neither works in producing happiness or a sense of lasting worth, however, because the position has not changed. "No matter what I do, I'm still NOT OK."

---

*Script Analysis is the method of uncovering the early decisions, made unconsciously, as to how life shall be lived. My reference to script and counterscript is general. Definitive studies of the origins and analysis of scripts are being conducted by a number of Transactional Analysts, notably Berne, Ernst, Groder, Karpman, and Steiner.

Once the position is uncovered and changed, the achievements and skills that have resulted from the counterscript can serve the person well when he builds a new and conscious life plan with the Adult.

I'm Not OK—You're Not OK

If all children who survive infancy initially conclude I'M NOT OK—YOU'RE OK, what happens to produce the second position, I'M NOT OK AND NEITHER ARE YOU. What happened to the YOU'RE OK? What happened to the source of stroking?

By the end of the first year something significant has happened to the child. He is walking. He no longer has to be picked up. If his mother is cold and nonstroking, if she only put up with him during the first year because she had to, then his learning to walk means that his "babying" days are over. The stroking ceases entirely. In addition punishments come harder and more often as he is able to climb out of his crib, as he gets into everything, and won't stay put. Even self-inflicted hurts come more frequently as his motility sends him tripping over obstacles and tumbling down stairs.

Life, which in the first year had some comforts, now has none. The stroking has disappeared. If this state of abandonment and difficulty continues without relief through the second year of life, the child concludes I'M NOT OK—YOU'RE NOT OK. In this position the Adult stops developing since one of its primary functions—getting strokes—is thwarted in that there is no source of stroking. A person in this position gives up. There is no hope. He simply *gets through* life and ultimately may end up in a mental institution in a state of extreme withdrawal, with regressive behavior which reflects a vague, archaic longing to get back to life as it was in the first year during which he received the only stroking he ever knew—as an infant who was held and fed.

It is hard to imagine anyone going through life without any stroking. Even with a nonstroking mother there most certainly appeared persons who were capable of caring for a person in this position and who, in fact, did stroke. However, once a position is decided, all experience is selectively interpreted to support it. If a person concludes YOU'RE NOT OK, it applies to all other people, and he rejects their stroking, genuine though it may be. He originally found some measure of integrity or sense in his early conclusion; therefore new experiences do not readily break it down. This is the deterministic nature of positions. Also, the individual in this position stops using his Adult with regard to his relationships with others. Therefore, even in treatment, it is difficult to reach his Adult, particularly in view of the fact that the therapist also occupies the category YOU'RE NOT OK.

There is one condition in which I'M NOT OK—YOU'RE NOT OK may be the initial position, rather than secondary to the first. This is the condition of the autistic child. The autistic child remains psychologically unborn. Infantile autism appears to be the response of the immature organism to catastrophic stress in an external world in which there is no stroking *which gets through to him.* The autistic child is one who, in the critical early weeks of life, did not feel himself to be rescued. It is as if he found "nobody out there" after his catastrophic expulsion into life.

Schopler* concludes there is a physiological factor which combines with insufficient stroking to produce the autistic child. The factor is thought to be a high stimulus barrier so that the stroking which is given does not register. He may not be totally deprived of stroking, but he may be deprived of his sensation of it, or an "accumulation" of his sensations of it. The infant is then seen by the parents as a nonresponsive child (he doesn't like to be held, he just lies there, he's different), and then even the stroking which has

---

*E. Schopler, "Early Infantile Autism and Receptor Processes," *Archives of General Psychiatry,* Vol. 13 (October 1965).

been given is withheld because "he doesn't like to be held." It is thought by some researchers that vigorous stroking (more than is given ordinarily) may have helped overcome the barrier.

I'm OK—You're Not OK

A child who is brutalized long enough by the parents he initially felt were OK will switch positions to the third, or criminal, position: I'M OK—YOU'RE NOT OK. There is OK-ness here, but where does it come from? Where is the source of stroking if YOU'RE NOT OK?

This is a difficult question considering that the position is decided in the second or third year of life. If a two-year-old concludes I'M OK, does this mean his OK is the product of "self-stroking," and, if so, how does a small child stroke himself?

I believe this self-stroking does in fact occur during the time that a little person is healing from major, painful injuries such as are inflicted on a youngster who has come to be known as "the battered child." This is the child who has been beaten so severely that bones and skin are broken. Anyone who has had a broken bone or massive bruises knows the pain. Common in battered children are extremely painful injuries such as broken ribs, smashed kidneys, and fractured skulls. How does the every-breath agony of broken ribs or the excruciating headache from blood in the spinal fluid feel to a toddler? Every hour five infants in this country receive injuries of this kind at the hands of their parents.

I believe that it is while this little individual is healing, in a sense "lying there licking his wounds," that he experiences a sense of comfort alone and by himself, if for no other reason than that his improvement is in such contrast to the gross pain he has just experienced. It is as if he senses, I'll be all right if you leave me alone. I'M OK by myself. As the brutal parents reappear, he may shrink in horror that it will happen again. You hurt me! You are NOT OK. I'M OK—YOU'RE NOT OK. The early history of many criminal psychopaths, who occupy this position, reveals this kind of gross physical abuse.

Such a little person has experienced brutality, but he has also experienced survival. What has happened can happen again. I did survive. I will survive. He refuses to give up. As he grows older he begins to strike back. He has seen toughness and knows how to be tough. He also has permission (in his Parent) to be tough and to be cruel. Hatred sustains him although he may learn to conceal it with a mask of measured politeness. Caryl Chessman said, "There is nothing that sustains you like hate; it is better to be anything than afraid."

For this child the I'M OK—YOU'RE NOT OK position is a life-saving decision. The tragedy, for himself and for society, is that he goes through life refusing to look inward. He is unable to be objective about his own complicity in what happens to him. It is always "their fault." It's "all them." Incorrigible criminals occupy this position. They are the persons "without a conscience" who are convinced that they are OK no matter what they do and that the total fault in every situation lies in others. This condition, which at one time was referred to as "moral imbecility," is actually a condition in which the person has shut out any incoming data that anyone is OK. For this reason treatment is difficult, since the therapist is NOT OK along with everyone else. The ultimate expression of this position is homicide, *felt* by the killer to be justifiable (in the same way that he felt justified in taking the position in the first place).

The person in the I'M OK—YOU'RE NOT OK position suffers from stroking deprivation. A stroke is only as good as the stroker. And there are no OK people. Therefore there are no OK strokes. Such a person may develop a retinue of "yes men" who praise and stroke him heavily. Yet he knows they are not authentic strokes because he has had to set them up himself, in the same way he had to produce his own stroking in the first place. The more they praise him the more despicable they become, until he finally rejects them all in favor of a new group of yes men. "Come close so I can let you have it" is an old recording. That's the way it was in the beginning.

I'm OK—You're OK

There is a fourth position, wherein lies our hope. It is the I'M OK— YOU'RE OK position. *There is a qualitative difference between the first three positions and the fourth position.* The first three are unconscious, having been made early in life. I'M NOT OK—YOU'RE OK came first and persists for most people throughout life. For certain extremely unfortunate children this position was changed to positions two and three. By the third year of life one of these positions is fixed in every person. The decision as to position is perhaps one of the first functions of the infant's Adult in the attempt to make sense out of life, so that a measure of predictability may be applied to the confusion of stimuli and feelings. These positions are arrived at on the basis of data from the Parent and Child. They are based on emotion or impressions without the benefit of external, modifying data.

The fourth position, I'M OK—YOU'RE OK, because it is a conscious and verbal decision, can include not only an infinitely greater amount of information about the individual and others, but also the incorporation of not-yet-experienced possibilities which exist in the abstractions of philosophy and religion. *The first three positions are based on feelings. The fourth is based on thought, faith, and the wager of action.* The first three have to do with *why.* The fourth has to do with *why not?* Our understanding of OK is not bound to our own personal experiences, because we can transcend them into an abstraction of ultimate purpose for all men.

We do not drift into a new position. It is a decision we make. In this respect it is like a conversion experience. We cannot decide on the fourth position without a great deal more information than most persons have available to them about the circumstances surrounding the original positions decided on so early in life. Fortunate are the children who are helped early in life to find they are OK by repeated exposure to situations in which they can prove, to themselves, their own worth and the worth of others.

Unfortunately, the most common position, shared by "successful" and "unsuccessful" persons alike, is the I'M NOT OK—YOU'RE OK position. The most common way of dealing with this position is by the playing of *games*. Berne defines a game as

> . . . an ongoing series of complementary ulterior transactions progressing to a well-defined, predictable outcome. Descriptively it is a recurring set of transactions, often repetitious, superficially plausible, with a concealed motivation; or, more colloquially, a series of moves with a snare, or "gimmick."*

I believe all games have their origin in the simple childhood game, easily observed in any group of three-year-olds: "Mine Is Better Than Yours." This game is played to bring a little momentary relief from the awful burden of the NOT OK. It is essential to keep in mind what the I'M NOT OK—YOU'RE OK position means to the three-year-old. I'M NOT OK means: I'm two feet tall, I'm helpless, I'm defenseless, I'm dirty, nothing I do is right, I'm clumsy, and I have no words with which to try to make you understand how it feels. YOU'RE OK means: You are six feet tall, you are powerful, you are always right, you have all the answers, you are smart, you have life or death control over me, and you can hit me and hurt me, and it's still OK.

Any relief to this unjust state of affairs is welcome to the child. A bigger dish of ice cream, pushing to get first in line, laughing at sister's mistakes, beating up little brother, kicking the cat, having more toys, all give momentary relief even though down the road is another disaster like a spanking, getting hit by little brother, being clawed by the cat, or finding someone who has more toys.

Grownups indulge in sophisticated variations of the "Mine Is Better" game. Some people achieve temporary relief by accumulating possessions, by living in a bigger, better house than the Joneses,

*E. Berne, *Games People Play* (New York: Grove Press, 1964), p. 48.

or even reveling in their modesty: I am humbler than you are. These maneuvers, which are based on what Adler called "guiding fictions," may provide a welcome relief even though down the road may be a disaster in the form of an oppressive mortgage or consumptive bills, which commit the person to a life of perpetual drudgery. In Chapter 7 games are explained in detail as a misery-producing "solution" which compounds the original misery and confirms the NOT OK.

The aim of this book is to establish that the only way people get well or become OK is to expose the childhood predicament underlying the first three positions and prove how current behavior perpetuates the positions.

Finally, it is essential to understand that I'M OK—YOU'RE OK is a *position and not a feeling*. The NOT OK recordings in the Child are not erased by a decision in the present. The task at hand is how to start a collection of recordings which play OK outcomes to transactions, successes in terms of correct probability estimating, successes in terms of integrated actions which make sense, which are programmed by the Adult, and not by the Parent or Child, successes based on an ethic which can be supported rationally. A man who has lived for many years by the decisions of an emancipated Adult has a great collection of such past experiences and can say with assurance, "I know this works." The reason I'M OK—YOU'RE OK works is that instant joy or tranquility is not expected.

One day a young divorcée in one of my groups complained angrily, "You and your damned OK bit! I went to a party last night and I decided to be just as nice as could be, and I decided everyone else there was OK. And I went up to this woman I know, and I said, 'Why don't you come over and have coffee with me sometime?' and she cut me down to two feet tall with 'Well, I would like to, but you know everybody doesn't have the time to sit around and gas all day the way you do.' It's for the birds . . . won't work!"

Personal or social storms are not going to subside immediately when we assume a new position. The Child wants immediate results—like instant coffee, thirty-second waffles, and immediate re-lief from acid indigestion. The Adult can comprehend that pa-tience and faith are required. We cannot guarantee instant OK feelings by the assuming of the I'M OK—YOU'RE OK position. We have to be sensitive to the presence of the old recordings; but we can choose to turn them off when they replay in a way that under-mines the faith we have in a new way to live, which, *in time,* will bring forth new results and new happiness in our living. The Adult also can recognize the Child responses in others and can choose not to respond in kind.

The change that this entails, and how change is possible, will be illustrated in the next chapter.

4

We Can Change

All men plume themselves on the improvement
of society, and no man improves.

—**Ralph Waldo Emerson**

To state that we have problems is not particularly helpful.
More to the point is the fact that most of our energy day after day
is used in decision making. Patients frequently say: I can't make up
my mind, tell me what to do, I'm afraid I'll make the wrong decision; or, in the face of inability to decide: I am always on the verge
of going to pieces, I hate myself, I can never seem to accomplish
anything, my life is a succession of failures.

Though these statements can be said to express problems,
they all originate in the difficulty that surrounds decision making.
The unsettling nature of indecisiveness is sometimes expressed in
the indiscriminate plea: Do something—anything—just do something. In treating patients we see two essential difficulties with decision making: (1) "I always make the wrong decision," which is
the expression of the person whose decisions and the activity
which follows generally turn out badly for him; and (2) "I keep
going over and over the same thing," which is the expression of the

person whose computer is cluttered with unfinished business or pending decisions.

The first step in solving either of these difficulties is to recognize that in each decision there are three sets of data that must be processed. The first set of data is in the Parent, the second in the Child, and the third in the Adult. Parent and Child data are dated. Adult data represent external reality as it exists in the present, together with a vast amount of data accumulated in the past, independent of the Parent and Child. Data from all three sources pour into the computer in response to a transactional stimulus. Which is going to respond—Parent, Adult, or Child? Perhaps the best way to explain this process is to illustrate.

We shall say that a middle-aged businessman, who has a reputation as a good father, a good husband, and a responsible citizen, has to make a decision whether to sign his name to a petition that will appear in the local newspaper. This petition supports a fair-housing bill to enable individuals of all races to live wherever their income allows. The request comes by telephone, and as soon as he hangs up he is aware of great discomfort, a churning in his stomach, and the feeling that a perfectly good day has been ruined.

He has a decision to make, and there is obviously a great amount of conflict about it. Where does the conflicting data come from?

One source is his Parent. Among the recordings that turn on are "Don't bring shame on the family"; "Don't stick your neck out"; "Why you?"; and "Your family and children must always come first!" These are overtones to an even more compelling recording, cut in his earliest years in his home in a Southern town, "You've got to keep them in their place." In fact there is a whole category of Parent data under the heading "nigger" which has been unavailable for inquiry. On this body of data the door was locked in early childhood by the firm directives, "Don't ask questions." "He's a nigger, that's why." "And don't let me hear you bring it up again!" (Even the "harmless" little rhymes like

"Eenie, meenie, minie, mo, Catch a nigger by the toe" orchestrate the theme.)

These early recordings, reinforced through the years by continuing parental dictates and by further evidence that the presence of Negroes can be the cause of trouble (for example, in Little Rock, Selma, Watts, and Detroit), are a powerful force bearing on this man's decision.

The power of this incoming data lies in its ability to reproduce fear in the Child. The "six-foot-tall" Parent is again at work on the "two-foot-tall" Child to make him conform. Thus the second set of data comes from the Child. These data are expressed mostly as feelings: fear, what will "they" say, what would happen if my daughter "married one," what will happen to the value of my property? There are realistic difficulties here, but the intensity of the feelings is not so much related to the realistic difficulties as to the original difficulty of the three-year-old child dependent upon his parents for security. This produces the churning stomach and the sweaty hands. The conflict can be so painful that the man heads for his liquor cabinet or some other evasive activity to "get his Parent off his back."

This would be a short-lived war were it not for another set of data, which also is being fed into the computer. This is the data which comes from reality and is in the particular domain of the Adult. A "simple" or "nonthinking" man isn't too much troubled by reality. He simply gives in to the Parent. His slogan is Peace (for the Child) at Any Price. The old ways are the best ways. It is human nature. History repeats itself. Let George do it.

It is only the man with an operating Adult who must take into account the seriousness of the threat of racial crisis even to his own well-being. Only his Adult can go for more data. Only his Adult can assess how the evil of slavery, or treating persons as things, has produced a humiliation and a hopelessness so devastating to many Negroes that it is expressed in Little Rock or Selma or Watts or Detroit. Only the Adult, like Lincoln, can say, "The dogmas of the quiet past are insufficient to the present struggle." Only the Adult can look objectively at *all* the data and proceed to look for more.

It is in this process of identifying and separating the three sets of data that we begin to bring order out of the chaos of feelings and indecision. Once separated, the three bodies of data can be examined by the Adult to see what is valid.

The questions our troubled businessman must ask in examining his Parent data are: Why did his parents believe these things? What was their Parent like? Why was their Child threatened? What was their ability or inability to examine their own P-A-C? Is what they believe true? Are white persons superior to black persons? Why? Why not? Is it wrong never to stick your neck out? Would an anti-discriminatory position necessarily "bring shame on the family"? Might it bring honor? Does he really place his family and children first if he does not contribute to a realistic solution of racial problems in his community? It might even be helpful to ask what his parents believe today in contrast to what they believed when his Parent was recorded.

His Adult must also examine the data coming from the Child. Why does he feel so threatened? Why is his stomach upset? Is there a realistic threat? Is his fear appropriate today? Or was it appropriate only when he was three years old? He may have realistic fears about rioting and violence. He could be killed; but he must differentiate between the fear produced by current events and the fear he felt at age three. The "age three" fear is far greater. At age three he cannot change reality. But at forty-three he can. He can take steps to change reality, and ultimately to change the circumstances which produce the realistic danger.

Understanding the "age three" fear is essential to freeing the Adult for processing new data. This is the fear—the archaic fear of the all-powerful Parent—which makes persons "prejudge," or which makes them prejudiced. A person who is prejudiced is like the little boy in Chapter 2 who accepts "cops are bad" as ultimate

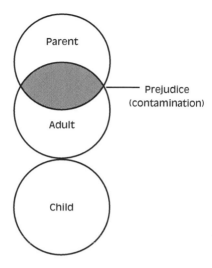

Fig. 8. Prejudice.

truth. He is afraid to do otherwise. This produces the contamination of the Adult (Figure 8), and this contamination allows prejudice, or unexamined Parent data, to be externalized as true.

To paraphrase Socrates, who held that "The unexamined life is not worth living," we may say the unexamined Parent is not worth basing one's life on: It might be wrong.

The Emancipated Adult

The goal of Transactional Analysis is to enable a person to have freedom of choice, the freedom to change at will, to change the responses to recurring and new stimuli. Much of this freedom is lost in early childhood, marking the onset, according to Kubie, of the "neurotic process." This process is one which is continually involved in solving archaic problems to the exclusion of dealing effectively with today's reality.

The roots of the neurotic process may lie in a . . . phenomenon of infancy and/or early childhood—the formation of stereotyped behavioral patterns or "fixations." These may include affective displays—for example, crying, screaming, night terrors; instinctual performance or non-performance—for example, eating, ruminating, vomiting, regurgitating, food refusals, compulsive food choices, patterns of excreting or retention; respiratory patterns such as breath holding, and so forth; or patterns of general actions—for example, tics, head bumping, rocking, sucking, and so forth.

Not one of these acts is in and of itself inherently abnormal. As long as it can change freely in response to changing external or internal cues, it remains normal. It is the loss of the freedom to change which marks the onset of the neurotic process. * [Italics mine.]

Restoration of the freedom to change is the goal of treatment. This freedom grows from knowing the truth about what is in the Parent and what is in the Child and how this data feeds into present-day transactions. It also requires the truth, or the *evidence,* about the world in which he lives. Such freedom requires the knowledge that everyone with whom one deals has a Parent, an Adult, and a Child. It requires persistent exploration not only into "knowable" areas but also into indeterminate areas, which can best be understood in terms of another function of the Adult, that of probability estimating. One of the realities of the human predicament is that we frequently have to make decisions before all the facts are in. This is true of any commitment. It is true of marriage. It is true of voting. It is true of signing a petition. It is true of the establishment of priorities. It is true of those values we embrace independently—that is, with the Adult.

The Child in us demands certainty. The Child wants to know the sun will come up every morning, that Mother will be there, that

*Lawrence Kubie, "Role of Polarity in Neurotic Process," *Frontiers of Clinical Psychiatry,* Vol. 3, No. 7 (April 1, 1966).

the "bad guy" will always get it in the end; but the Adult can accept
the fact that there is not always certainty. Philosopher Elton Trueblood states:

> The fact that we do not have absolute certainty in regard to
> any human conclusions does not mean that the task of inquiry is
> fruitless. We must, it is true, always proceed on the basis of
> probability, but to have probability is to have something. What
> we seek in any realm of human thought is not absolute certainty,
> for that is denied us as men, but rather the more modest path of
> those who find dependable ways of discerning different degrees
> of probability.*

This is in the explorative realm of philosophy and religion, into
which we will look further in Chapter 12, "P-A-C and Moral Values."

The Adult in our businessman, confronted with the housing
petition, can admit that the outcome of his signing is uncertain. If
he signs, he may be ridiculed. If his I'M OK—YOU'RE OK position includes all persons, regardless of race or religion, he may be attacked
by prejudiced persons who are in a position to hurt his income, his
membership in the golf club, or his relationship with his wife. But
he also can weigh the possibility that his contribution to a solution
to racial unrest in his community may lead to a significant reduction in the problem. In the long run it may bring stroking to his
Child in the form of a reputation as a man with integrity and
courage.

When the Parent or the Child dominates, the outcome is predictable. This is one of the essential characteristics of games. There
is a certain security in games. They may always turn out painfully,
but it is a pain that the player has learned to handle. When the Adult
is in charge of the transaction, the outcome is not always predictable. There is the possibility of failure, but there is also the possibility of success. Most important, there is the possibility of change.

*Elton Trueblood, *General Philosophy* (New York: Harper, 1963).

What Makes People Want to Change?

Three things make people want to change. One is that they hurt sufficiently. They have beat their heads against the same wall so long that they decide they have had enough. They have invested in the same slot machines without a pay-off for so long that they finally are willing either to stop playing or to move on to others. Their migraines hurt. Their ulcers bleed. They are alcoholic. They have hit the bottom. They beg for relief. They want to change.

Another thing that makes people want to change is a slow type of despair called ennui, or boredom. This is what the person has who goes through life saying, "So what?" until he finally asks the ultimate big "*So What?*" He is ready to change.

A third thing that makes people want to change is the sudden discovery that they can. This has been an observable effect of Transactional Analysis. Many people who have shown no particular desire to change have been exposed to Transactional Analysis, through lectures or by hearing about it from someone else. This knowledge has produced an excitement about new possibilities, which has led to their further inquiry and a growing desire to change. There also is the type of patient who, although suffering from disabling symptoms, still does not really want to change. His treatment contract reads, "I'll promise to let you help me if I don't have to get well." This negative attitude changes, however, as the patient begins to see that there is indeed another way to live. A working knowledge of P-A-C makes it possible for the Adult to explore new and exciting frontiers of life, a desire which has been there all along but has been buried under the burden of the NOT OK.

Does Man Have a Free Will?

Can man really change if he wants to, and if he can, is even his changing a product of past conditioning? Does man have a will?

One of the most difficult problems of the Freudian position is the problem of determinism versus freedom. Freud and most behaviorists have held that the cause-and-effect phenomenon seen in all the universe also holds true for human beings, that whatever happens today can theoretically be understood in terms of what has happened in the past. If a man today murders another man, we are accustomed by Freudian orientation to look into his past to find out why. The assumption is that there must be a cause or causes, and that the cause or causes lie somewhere in the past. The pure determinist holds that man's behavior is not free and is only a product of his past. The inevitable conclusion is that man is not responsible for what he does; that, in fact, he does not have a free will. The philosophical conflict is seen most dramatically in the courts. The judicial position is that man is responsible. The deterministic position, which underlies much psychiatric testimony, is that man is not responsible by virtue of the events of his past.

We cannot deny the reality of cause and effect. If we hit a billiard ball and it strikes several more, which then are impelled to strike other billiard balls in turn, we must accept the demonstration of the chain sequence of cause and effect. The monistic principle holds that laws of the same kind operate in all nature. Yet history demonstrates that while billiard balls have become nothing more than what they are as they are caught in the cause-and-effect drama, human beings have become more than what they were.

Will Durant has commented on how nineteenth-century French philosopher Henri Bergson relentlessly pressed the issue of determinism to absurdity:

> Finally, was determinism any more intelligible than free will? If the present moment contains no living and creative choice, and is totally and mechanically the product of the matter and motion of the moment before, then so was that moment the mechanical effect of the moment that preceded it, and that again of the one before . . . and so on, until we arrive at the primeval nebula as the total cause of every later event, of every line of Shake-

speare's plays, and every suffering of his soul; so that the som-
bre rhetoric of Hamlet and Othello, of Macbeth and Lear, in
every clause and every phrase, was written far off there in the
distant skies and the distant aeons, by the structure and content
of that legendary cloud. What a draft upon credulity. . . . There
was matter enough for rebellion here; and if Bergson rose so
rapidly to fame it was because he had the courage to doubt
where all the doubters piously believed.*

The answer lies not in refuting the cause-and-effect nature of
the universe or of man's behavior but in looking elsewhere than in
the past for cause. Man does what he does for certain reasons, but
those reasons do not all lie in the past. In a television interview I
was asked my opinion as to why Charles Whitman climbed a tower
at the University of Texas and shot scores of people on the ground
below. After a recount of a number of possible reasons I was asked,
"But why do some people do a thing like this and others do not?"
This is a valid question. If our position is that we simply don't
know enough about the past history of an individual, then we still
hold to the position that somewhere "back there" lies an answer.

There is an essential difference, however, between a man and a
billiard ball. Man, through thought, is able to look to the future. He
is influenced by another type of causal order which Charles
Harteshorne calls "creative causation."† Elton Trueblood elabo-
rates this point by suggesting that causes for human behavior lie
not only in the past but in man's ability to contemplate the future,
or estimate probabilities:

> The human mind . . . operates to a large extent by reference to
> final causes. This is so obvious that it might seem impossible to
> neglect it, yet it is neglected by everyone who denies freedom in

*Will Durant, *The Story of Philosophy* (New York: Simon and Schuster 1933), pp.
337–338.
†See "Causal Necessities, an Alternative to Hume," *The Philosophical Review* 63
(1954), pp. 479–499.

employing the billiard ball analogy of causation. Of course, the billiard ball moves primarily by efficient causation, but man operates in a totally different way. Man is a creature whose present is constantly being dominated by reference to the nonexistent, but nevertheless potent, future. What is *not*, influences what *is*. I have a hard problem but the outcome is not merely the result of a mechanical combination of forces, which is true of a physical body; instead I think, and most of my thought is concerned with what might be produced, provided certain steps could be taken.*

Ortega defines man as "a being which consists not so much in what it is as in what it is going to be."† Trueblood points out

> . . . it is not enough to say that the outcome is determined even by one's previous character, for the reality in which we share is such that genuine novelty can emerge in the very act of thinking. Thinking, as we actually experience it daily, is not merely awareness of action, as it is in all epiphenomenalist doctrine, but is a true and creative *cause*. Something happens, when a man thinks, which would not have occurred otherwise. This is what is meant by self-causation as a genuine third possibility in our familiar dilemma. ‡

Thus we see the Adult as the place where the action is, where hope resides, and where change is possible.

*Trueblood, *General Philosophy*.
†J. Ortega y Gasset, *What Is Philosophy?* (New York: Norton, 1960).
‡Trueblood, *General Philosophy*.

5

Analyzing the Transaction

I do not understand my own actions.

—Saint Paul

Now that we have developed a language, we come to the central technique: using that language to *analyze a transaction*. The transaction consists of a stimulus by one person and a response by another, which response in turn becomes a new stimulus for the other person to respond to. The purpose of the analysis is to discover which part of each person—Parent, Adult, or Child—is originating each stimulus and response.

There are many clues to help identify stimulus and response as Parent, Adult, or Child. These include not only the words used but also the tone of voice, body gestures, and facial expressions. The more skillful we become in picking up these clues, the more data we acquire in Transactional Analysis. We do not have to dig deep into anecdotal material in the past to discover what is recorded in Parent, Adult, and Child. We reveal ourselves today.

The following is a list of physical and verbal clues for each state.

PARENT CLUES—PHYSICAL

Furrowed brow, pursed lips, the pointing index finger, head-wagging, the "horrified look," foot-tapping, hands on hips, arms folded across chest, wringing hands, tongue-clucking, sighing, patting another on the head. These are typical Parent gestures. However, there may be other Parent gestures peculiar to one's own Parent. For instance, if your father had a habit of clearing his throat and looking skyward each time he was to make a pronouncement about your bad behavior, this mannerism undoubtedly would be apparent as your own prelude to a Parent statement, even though this might not be generally seen as Parent in most people. Also, there are cultural differences. For instance, in the United States people exhale as they sigh, whereas in Sweden they inhale as they sigh.

PARENT CLUES—VERBAL

I am going to put a stop to this *once and for all;* I can't for the life of me . . . ; Now always remember . . . ; ("always" and "never" are *almost always* Parent words, which reveal the limitations of an archaic system closed to new data); How many times have I told you? If I were you . . .

Many evaluative words, whether critical or supportive, *may* identify the Parent inasmuch as they make a judgment about another, based not on Adult evaluation but on *automatic,* archaic responses. Examples of these kinds of words are: stupid, naughty, ridiculous, disgusting, shocking, asinine, lazy, nonsense, absurd, poor thing, poor dear, no! no!, sonny, honey (as from a solicitous saleslady), How dare you?, cute, there there, Now what?, Not again! It is important to keep in mind that these words are *clues,* and are not conclusive. The Adult may decide after serious deliberation that, on the basis of an Adult ethical system, certain things *are* stupid, ridiculous, disgusting, and shocking. Two words, "should" and "ought," frequently are giveaways to the Parent state, but as I contend in Chapter 12, "should" and "ought" can also be Adult words. It is the automatic, archaic, *unthinking* use of these words which signals the activation of the Parent. The use of these words, together with body gestures and the context of the transaction, helps us identify the Parent.

CHILD CLUES—PHYSICAL

Since the Child's earliest responses to the external world were non-verbal, the most readily apparent Child clues are seen in physical expressions. Any of the following signal the involvement of the Child in a transaction: tears; the quivering lip; pouting; temper tantrums; the high-pitched, whining voice; rolling eyes; shrugging shoulders; downcast eyes; teasing; delight; laughter; hand-raising for permission to speak; nail-biting; nose-thumbing; squirming; and giggling.

CHILD CLUES—VERBAL

Many words, in addition to baby talk, identify the Child: I wish, I want, I dunno, I gonna, I don't care, I guess, when I grow up, bigger, biggest, better, best (many superlatives originate in the Child as "playing pieces" in the "Mine Is Better" game). In the same spirit as "Look, Ma, no hands," they are stated to impress the Parent and to overcome the NOT OK.

There is another grouping of words which are spoken continually by little children. However, these words are not clues to the Child, but rather to the Adult operating in the little person. These words are why, what, where, who, when, and how.

ADULT CLUES—PHYSICAL

What does the Adult look like? If we turn off the video on the Parent and Child tapes, what will come through on the face? Will it be blank? Benign? Dull? Insipid? Ernst* contends that the blank face does not mean an Adult face. He observes that listening with the Adult is identified by continual movement—of the face, the eyes, the body—with an eyeblink every three to five seconds. Nonmovement signifies non-listening. The Adult face is straightforward, says Ernst. If the head is tilted, the person is listening with an angle in mind. The Adult also allows the curious, excited Child to show its face.

ADULT CLUES—VERBAL

As stated before, the basic vocabulary of the Adult consists of why, what, where, when, who, and how. Other words are: how much, in what way, comparative, true, false, probable, possible, unknown, objective, I think, I see, it is my opinion, etc. These words all indicate Adult data processing. In the phrase "it is my opinion," the opinion may be derived from the Parent, but the statement is Adult in that it is identified as an opinion and not as fact. "It is my opinion that high school students should vote" is not the same as the statement "High school students should vote."

With these clues to assist us, we can begin to identify Parent, Adult, and Child in transactions involving ourselves and others.

Any social situation abounds with examples of every conceivable type of transaction. On a fall day some years ago I was riding a Greyhound bus to Berkeley and made a note of a number of transactions. The first was a Parent-Parent exchange (Figure 9) between two cheerless ladies, seated side by side, across from me. They were developing a rather extensive philosophy around the point of whether or not the bus would get to Berkeley on time. With great knowing, sympathetic nods of the head they produced a long exchange which began with the following transactions:

*F. Ernst, lecture on "Listening" delivered at the Institute for Transactional Analysis, Sacramento, California, Oct. 18, 1967.

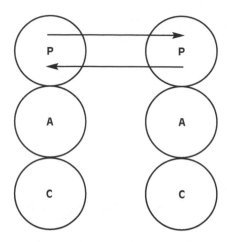

Fig. 9. Parent-Parent transaction.

LADY 1: (Looks at her watch, winds it, mumbles, catches the eye of the lady next to her, sighs wearily.)

LADY 2: (Sighs back, shifts uncomfortably, looks at her watch.)

LADY 1: Looks like we're going to be late again.

LADY 2: Never fails.

LADY 1: You ever see a bus on time—ever?

LADY 2: Never have.

LADY 1: Just like I was saying to Herbert this morning—you just don't get service any more like you used to.

LADY 2: You're absolutely right. It's a sign of the times.

LADY 1: It costs you, though. You can count on that!

These transactions are Parent-Parent in that they proceed without the benefit of reality data and are the same kind of judgmental exchange these ladies, as children, overheard between their mommies and aunties over the vicissitudes of riding streetcars. Lady 1 and Lady 2 enjoyed recounting the "awfuls" more than they would have enjoyed getting the facts. This is because of the good feeling that comes from blaming and finding fault. When we blame and find fault, we replay the early blaming and fault-finding which is recorded in the Parent, and this makes us feel OK, because the Parent is OK, and we are coming on Parent. Finding someone to agree with you, and play the game, produces a feeling well-nigh omnipotent.

Lady 1 made the first move. Lady 2 could have stopped the game had she responded, at any point, with an Adult statement to any of Lady 1's statements:

LADY 1: (Looks at her watch, winds it, mumbles, catches the eye of the lady next to her, sighs wearily.)

Adult Response Possibilities

1. Nonacknowledgment of sigh, by looking away.
2. A simple smile.
3. (If Lady 1 were sufficiently distressed): "Are you all right?"

LADY 1: Looks like we're going to be late again.

Adult Response Possibilities

1. What time is it now?
2. This bus is usually on time.
3. Have you been late before?
4. I'll ask.

LADY 1: You ever see a bus on time—ever?

Adult Response Possibilities

1. Yes.
2. I don't usually ride the bus.
3. I've never thought about it.

LADY 1: Just like I was saying to Herbert this morning—you just don't get service any more like you used to.

Adult Response Possibilities

1. I can't agree with that.
2. What kind of service do you mean?
3. The standard of living is as high as ever, the way I see it.
4. I can't complain.

These alternative responses would have been Adult, but not complementary. Someone who is enjoying a game of "Ain't It Awful" does not welcome the intrusion of facts. If the neighbor girls enjoy an every-morning session of "Husbands Are Stupid," they will not welcome the new girl who announces brightly that her husband is a jewel.

This brings us to the first rule of communication in Transactional Analysis. When stimulus and response on the P-A-C transactional diagram make parallel lines, the transaction is complementary and can go on indefinitely. It does not matter which way the vectors go (Parent-Parent, Adult-Adult, Child-Child, Parent-Child, Child-Adult) if they are parallel. Lady 1 and Lady 2 did not make sense in terms of the facts, but their dialogue was complementary and continued for about ten minutes.

The "enjoyable misery" of the two lady passengers came to an end when the man in front of them asked the driver if they would be in Berkeley on time. The driver said, "Yes—at 11:15." This, too, was a complementary transaction between the man and the driver, Adult-Adult (Figure 10). It was a direct answer to a direct request for information. There was no Parent component (How are our chances of getting to Berkeley on time for a change?) and no Child component (I don't know why I always manage to get on the slowest bus). It was a dispassionate exchange. This kind of transaction gets the facts.

Behind the two women were two other people, whose activity illustrates another type of transaction, Child-Child. One was a

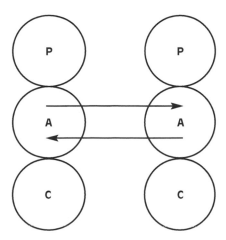

Fig. 10. Adult-Adult transaction.

fuzzy-faced, surly-looking boy with unkempt hair, who was wearing dusty, black trousers matched by a black-leather jacket. The other adolescent was dressed similarly and wore a look of forced dissipation. Both were engrossed in reading the same paperbacked book, *Secrets of the Torture Cult.* Had two priests been poring over the same book one might have assumed they were looking for Adult data about this strange subject; but from observing these two adolescent boys one was more likely to assume that this was a Child-Child transaction, involving somewhat the same cruel pleasure two five-year-old boys might find in discovering how to pull the wings off flies. Let us assume the adolescents acted on their new knowledge and found a way to torture someone as outlined in their text. There would be no Adult input (no understanding of consequences) and no Parent input ("It's horrible to do something like that"). Even if the transaction turned out badly for them (the arrival of the police—or of a mother in the case of the five-year-olds pulling wings off flies), the two persons involved in the transaction itself would have been in agreement. Therefore, it is complementary, Child-Child (Figure 11).

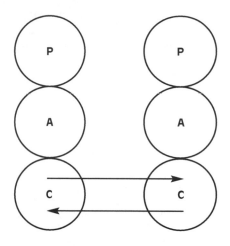

Fig. 11.　Child-Child transaction.

ADDITIONAL ILLUSTRATIONS OF COMPLEMENTARY TRANSACTIONS

Parent-Parent Transactions (See Figure 9)

STIMULUS:　Her duty is home with the children.

RESPONSE:　She obviously has no sense of duty.

STIMULUS:　It is disgusting the way taxes keep going up to feed all these no-goods at the public trough.

RESPONSE:　Where will it all end?

STIMULUS:　Kids nowadays are lazy.

RESPONSE:　It's a sign of the times.

STIMULUS:　I'm going to get to the bottom of this once and for all.

RESPONSE:　You should! You have to nip this kind of thing in the bud.

STIMULUS: Illegitimate, you know.

RESPONSE: Oh, *that* explains it.

STIMULUS: John fired? How *dare* they do such a thing?

RESPONSE: There, there, honey. I don't know why he worked for that stupid company in the first place.

STIMULUS: She married him for his money.

RESPONSE: Well, that's *all* she got.

STIMULUS: You can never trust one of those people.

RESPONSE: Exactly! Their kind are all alike.

Adult-Adult Transactions (See Figure 10)

STIMULUS: What time is it?

RESPONSE: I have 4:30.

STIMULUS: That is a good-looking suit.

RESPONSE: Thank you.

STIMULUS: This new ink dries very quickly.

RESPONSE: Is it more expensive than the other kind?

STIMULUS: Please pass the butter.

RESPONSE: There you are.

STIMULUS: What smells so good, dear?

RESPONSE: Cinnamon rolls in the oven . . . almost ready!

STIMULUS: I don't know what to do. I can't decide what's right.

RESPONSE: I don't think you should try to make a decision when you are so weary. Why don't you go to bed and we'll talk about it in the morning?

STIMULUS: Looks like rain.

RESPONSE: That's the forecast.

STIMULUS: Public relations is a function of management.

RESPONSE: You mean it can't be arranged through an agency?

STIMULUS: The *Lurline* sails at 1 o'clock Friday.

RESPONSE: What time do we have to be there?

STIMULUS: John has seemed worried lately.

RESPONSE: Why don't we have him over for dinner?

STIMULUS: I am tired.

RESPONSE: Let's go to bed.

STIMULUS: I see where federal taxes are going up again next year.

RESPONSE: Well, that's not good news. But if we're going to keep spending we've got to get the money somewhere.

Child-Child Transactions (See Figure 11)

It becomes readily apparent that there are very few game-free complementary Child-Child transactions. This is because the Child is a get-stroke rather than a give-stroke creature. People have transactions to get stroking. Bertrand Russell said: "One can't think hard from a mere sense of duty. I need little successes from time to time to keep . . . a source of energy."* Without Adult involvement in the transaction, no stroking accrues to anyone, and the relationship becomes uncomplementary, or dies of boredom.

A clear social example of this phenomenon is the hippie move-

*B. Russell, *The Autobiography of Bertrand Russell* (Boston: Little, Brown, 1967).

ment. The flower children extolled a life of Child-Child transactions. Yet the dreadful truth began to become apparent: It's no fun to do *your* thing if everybody else is only interested in doing *his* thing. In cutting off the Establishment they cut off the Parent (disapproval) and the Adult ("banal" reality); but, having cut off the disapproval, they found they had also cut off the source of praise. (A couple of four-year-olds may decide to run away from home, but give up the idea when they think it would be nice to have an ice-cream cone, and that takes mommies.) The flower children looked to each other for strokes but these became more and more impersonal and meaningless: Boy to girl: "Of course I love you. I love everybody!" Life thus began to settle down into more and more primitive means of stroking, such as fantasy stroking (withdrawal with drugs) and continual sexual activity. Sex can be solely a Child-Child activity inasmuch as the sexual urge is a genetic recording in the Child, as are all primary biological urges. The most pleasurable sex is more, however, in that there is an Adult component of considerateness, gentleness, and responsibility for the feelings of another. Not all hippies are devoid of these values, just as not all hippies are devoid of a Parent and Adult. Many, however, live on a self-seeking basis and, in a sense, *use* each other for sensory stimulation.

Happy hippie relationships, or childhood friendships which are full of fun, will be found to contain not wholly Child-Child transactions but Adult data-processing and Parent values as well. For example, two little girls playing:

GIRL 1 (CHILD): I'll be the mamma and you be the little girl.

GIRL 2 (CHILD): I always have to be the little girl.

GIRL 1 (ADULT): Well, let's take turns; you be the mamma first, and then next time I'll be the mamma.

This exchange is not Child-Child because of the Adult input (problem solving) apparent in the last statement.

Also, many of the transactions of small children are Adult-Adult, although they may seem "childish" because of data deficiency:

LITTLE GIRL: Emergency, Emergency! Buzzy [the cat] lost a tooth.

SISTER: Does the tooth fairy bring money to cats?

Both stimulus and response are Adult—valid statements on the basis of the data at hand. Good data processing; wrong data! Complementary Child-Child transactions can more readily be observed in what persons *do* together than in what they say to each other—as is true of very small children. A couple holding on to each other for dear life and screaming at the top of their lungs in the middle of a roller coaster ride are having a Child-Child transaction. Tagliavini and Tassinari singing the Act III duet from *Mefistofele* could be said to be having an intense Child-Child transaction. Grandma and Grandpa walking barefoot on the beach could be said to be having a Child-Child transaction. Yet the Adult made the arrangements for these happy experiences. It took money to ride the roller coaster. Tagliavini and Tassinari trained for years in order to experience the ecstasy of singing. Grandma and Grandpa share the joys of togetherness made possible by a lifetime of give-and-take. A relationship between people cannot last very long without the Adult. Thus we may say that complementary Child-Child transactions exist with the permission and supervision of the Adult. When the Adult is not around, the Child gets snarled up in crossed transactions, which will be described later in this chapter.

Parent-Child Transactions

Another type of complementary transaction is one between Parent and Child (Figure 12). The husband (Child) is sick, has a fever, and wants attention. The wife (Parent) knows how ill he feels and is willing to mother him. This can go on in a satisfactory way indefinitely

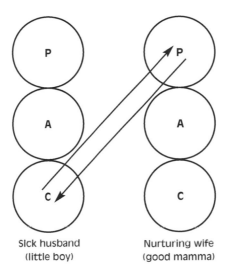

Sick husband   Nurturing wife
(little boy)   (good mamma)

Fig. 12. Child-Parent transaction.

as long as the wife is willing to be mothering. Some marriages are of this nature. If a husband wants to play "little boy" and his wife is willing to be parental, take the responsibility for everything, and look after him, this *can* be a satisfying marriage so long as neither wishes to change roles. If one or the other tires of the arrangement, the parallel relationship is disturbed, and trouble begins.

In Figure 13 we diagram a complementary transaction between George F. Babbitt (Parent) and Mrs. Babbitt (Child):

BABBITT (looking at the newspaper): "Lots of news. Terrible big tornado in the South. Hard luck, all right. But this, say, this is corking! Beginning of the end for those fellows! New York Assembly has passed some bills that ought to completely outlaw the socialists! And there's an elevator-runners' strike in New York and a lot of college boys are taking their places. That's the stuff! And a mass-meeting in Birmingham's demanded that this Mick agitator, this fellow DeValera, be deported.

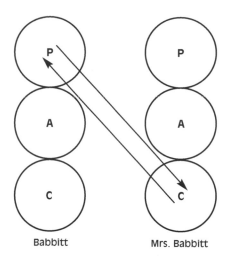

Fig. 13. Parent-Child transaction.

Dead right, by golly! All these agitators paid with German gold anyway. And we got no business interfering with the Irish or any other foreign government. Keep our hands strictly off. And there's another well-authenticated rumor from Russia that Lenin is dead. That's fine. It's beyond me why we don't just step in there and kick those Bolshevik cusses out."

MRS. BABBITT: "That's so."*

Child-Adult Transactions

Another type of complementary transaction is one between Child and Adult (Figure 14). A person in the grip of NOT OK feelings may reach out to another person for realistic reassurances. A husband may fear an upcoming business encounter, which a promotion de-

*Sinclair Lewis, "Babbitt," *Major American Writers,* ed. H. M. Jones and E. E. Leisy (New York: Harcourt, Brace, 1945), pp. 1,736.

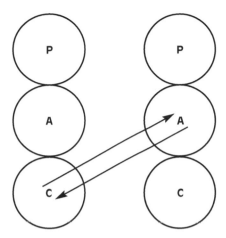

Fig. 14. Child-Adult transaction.

pends on. Even though he is qualified in every respect, he has an overload of Child data coming into his computer: I'm not going to make it! So he says to his wife, "I'm not going to make it," hoping for her recount of the reality reasons why he can make it if he doesn't let his NOT OK Child ruin his chances. He knows she has a good Adult and "borrows it" when his own is impaired. Her response is different from a Parent response, which might be reassuring even if reality data were not present or which might simply deny the Child feelings: "Of course you'll make it; don't be stupid!"

Adult-Parent Transactions

Another type of complementary transaction is Adult-Parent (Figure 15) and is represented by a man who wants to quit smoking. He has adequate Adult data as to why this is important to his health. Despite this, he asks his wife to play the Parent, to destroy his cigarettes when she finds them, to "come on strong" if he lights

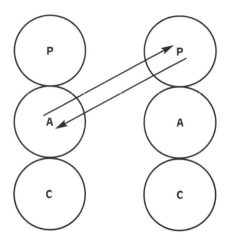

Fig. 15. Adult-Parent transaction.

one. This transaction has very good game possibilities. As soon as he turns the responsibility over to his wife's Parent, the husband can be a naughty little boy and play "If It Weren't for You I Could" or "Try and Catch Me."

Uncomplementary, or Crossed, Transactions

The kind of transaction that causes trouble is the *crossed transaction* (Figure 16). Berne's classical example is the transaction between husband and wife where husband asks: "Dear, where are my cuff links?" (An Adult stimulus, seeking information.) A complementary response by wife would be, "In your top left dresser drawer," or "I haven't seen them but I'll help you look."

However, if Dear has had a rough day and has saved up a quantity of "hurts" and "mads" and she bellows, "Where you left them!" the result is a crossed transaction. The stimulus was Adult but the wife turned the response over to the Parent.

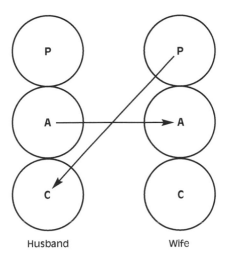

P   P

A ——————→ A

C   C

Husband   Wife

Fig. 16. Crossed transaction.

This brings us to the second rule of communication in Transactional Analysis. When stimulus and response cross on the P-A-C *transactional diagram, communication stops.* The husband and wife can't talk about cuff links anymore; they first have to settle why he never puts anything away. Had her response been Child ("Why do you always have to yell at me?") (Figure 17), the same impasse would have developed. These crossed transactions can set off a whole series of noisy exchanges which end with a bang somewhere in the purple outer reaches of "So's your old man!" Repetitious patterns of this type of exchange are what constitute games such as "It's All You," "If It Weren't for You I Could," "Uproar," and "Now I've Got You, You S.O.B.," which will be further elaborated in Chapter 7.

The origin of the non-Adult responses is in the NOT OK position of the Child. A person dominated by the NOT OK "reads into" comments that which is not there: "Where did you get the steaks?" "What's wrong with them?"; I *love* your new hair-do!" "You never did like it long"; "I hear you're moving." "We can't really afford it

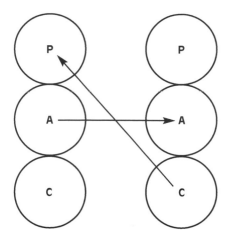

Fig. 17. Crossed transaction.

but this neighborhood is getting run down"; "Pass the potatoes, dear." "And you call *me* fat" As one of my patients said, "My husband says I could read something into a cookbook."

ADDITIONAL ILLUSTRATIONS OF CROSSED TRANSACTIONS

PATIENT (A): I would like to work in a hospital like this.

NURSE(P): You can't cope with your own problems. (Figure 18)

MOTHER (P): Go pick up your room.

DAUGHTER (P): You can't tell me what to do. You're not the boss around here. Dad's the boss! (Figure 19)

THERAPIST(A): What is your principal hang-up in life?

PATIENT (C): Red tape, red tape (pounding table), damn it, *red tape!* (Figure 20)

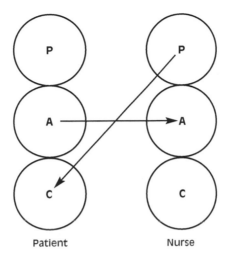

Fig. 18.

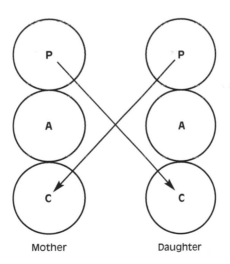

Fig. 19.

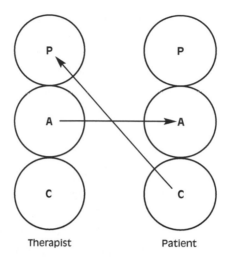

Fig. 20.

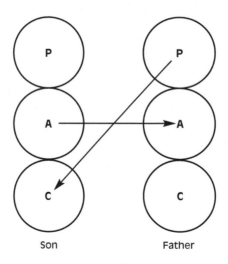

Fig. 21.

SON (A): I have to finish a report tonight that's due tomorrow.

FATHER (P): Why do you always leave things to the last minute? (Figure 21)

MAN (A), standing with friend: We were trying to get this gas cap unlocked and dropped the key behind the bumper. Could you help us get it out?

SERVICE STATION ATTENDANT (P): Who did it? (Figure 22)

LITTLE GIRL (A): Dirty shirts are warm.

MOTHER (P): Go take a bath. (Figure 23)

AOLESCENT GIRL (P): Well, frankly, my father likes Palm Springs best.

FRIEND (P): Our family tries to avoid the tourist places. (Figure 24)

LITTLE GIRL(C): I hate soup. I'm not going to eat it. You cook icky.

MOTHER (C): I'm just going to leave and then you can cook your own icky food. (Figure 25)

LITTLE BOY (C): My daddy has a million dollars.

LITTLE GIRL (C): That's nothing. My daddy has "finnegan" dollars. ("Finnegan" was this four-year-old's way of saying "infinity.") (Figure 26)

BABBITT'S DAUGHTER, VERONA (A): "I know, but—oh, I want to contribute—I wish I were working in a settlement house. I wonder if I could get one of the department stores to let me put in a welfare-department with a nice rest-room and chintzes and wicker chairs and so on and so forth. Or I could—"

BABBITT (P): "Now you look here! The first thing you got to understand is that all this uplift and flipflop and settlement work

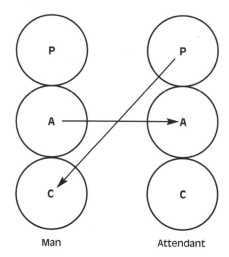

Man   Attendant

Fig. 22.

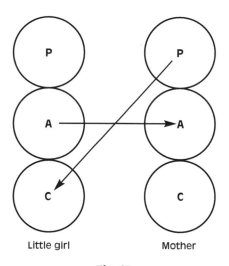

Little girl   Mother

Fig. 23.

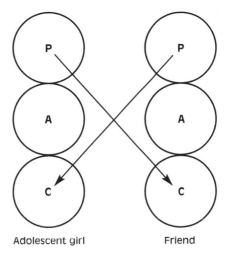

Fig. 24.

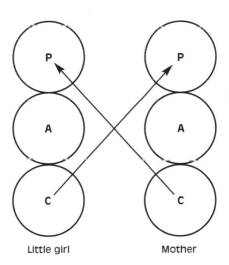

Fig. 25.

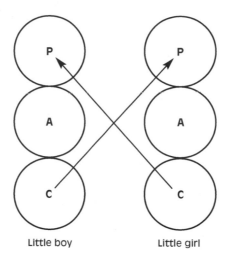

Little boy   Little girl

Fig. 26.

and recreation is nothing in God's world but the entering wedge for socialism. The sooner a man learns he isn't going to be coddled, and he needn't expect a lot of free grub, and, uh, all these free classes and flipflop and doodads for his kids unless he earns 'em, why, the sooner he'll get on the job and produce—produce—produce! That's what the country needs, and not all this fancy stuff that just enfeebles the will-power of the working man and gives his kids a lot of notions about their class. And you—if you'd tend to business instead of fooling and fussing—All the time! When I was a young man I made up my mind what I wanted to do, and stuck to it through thick and thin, and that's why I'm where I am today." (Figure 27)*

Parent responses, like Babbitt's, still stem from the NOT OK in the Child. He felt that his children did not appreciate him, that

*Lewis, *Babbitt.*

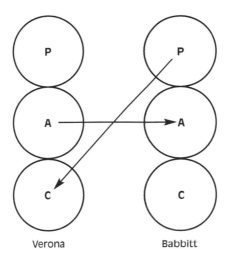

P   P

A ————→ A

C   C

Verona   Babbitt

Fig. 27.

they did not comprehend how hard he had struggled; he still felt NOT OK around those who had more than he did. If he had let his Child come on straight, he might have wept. So he took the safer course and turned the transaction over to the Parent, wherein resided self-righteousness, correctness, and "all the answers."

The person whose NOT OK Child is always activated cannot get on with transactions which will advance his dealing with reality because he is continually concerned with unfinished business having to do with a past reality. He can't accept a compliment gracefully because he doesn't think he deserves it, and there must be a barb in it somewhere. He is involved in a continuous attempt to maintain the integrity of the position that was established in the situation of childhood. The person who always comes on Child is really saying; "Look at me, I'm NOT OK." The person who is always coming on Parent is really saying, "Look at you, you're NOT OK (and that makes me feel better)." Both maneuvers are an expression of the NOT OK position and each contributes to the prolongation of despair.

The NOT OK position is not solely expressed in the response. It also can be found in the stimulus. Husband says to wife, "Where did you hide the can opener?" The main stimulus is Adult in that it seeks objective information. But there is a secondary communication in the word *hide*. (Your housekeeping is a mystery to me. We'd go broke if I were as disorganized as you. If I could once, just once, find something where it belongs!) This is Parent. It is a thinly veiled criticism. This stimulates a *duplex transaction* (Figure 28).

The progress of this transaction depends on which stimulus the wife wishes to respond to. If she wants to keep things amiable and feels OK enough not to have been threatened she may respond, "I hid it next to the tablespoons, darling." This is complementary in that she gives him the information he desires and also acknowledges good-naturedly his "aside" about her housekeeping. If her Adult computes that it is important to her marriage to do something about her husband's gentle suggestion, she may take the hint and become more organized. With her Adult handling the transaction, she can.

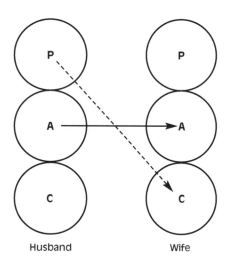

Fig. 28.

However, if her NOT OK Child is hooked, her primary response will be to the word *hide,* and she may respond along the lines of, "So what's the matter with you—you blind or something?" And there endeth the quest for the can opener while they wrangle over each other's merits and demerits in the area of organization, blindness, stupidity, etc. His beer is still unopened, and a game of "uproar" is well along.

Some transactions of this nature can involve stimulus and response at all levels: A man comes home and writes "I love you" in the dust on the coffee table. The Adult is in command of the situation, although both his Parent and Child are involved (Figure 29). The Parent says, "Why don't you ever clean this place up?" The Child says, "Please don't get mad at me if I criticize you." The Adult takes charge, however, on the basis that to be loving is important to my marriage, so I won't let my Parent or my Child come on straight. If I tell her I love her she won't get mad at me, but perhaps she'll get the idea that it is important, after all, for a man in my position to have a home that looks nice.

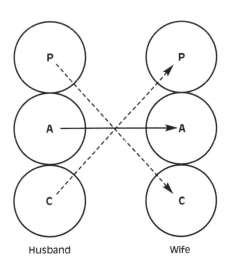

Fig. 29.

updated

This can turn into a complementary transaction if the wife is OK enough to take a little constructive criticism. The outcome would be happy if she shined up the house, met her husband at the door with a tall, cool drink, and told him what a sweet sentimental, imaginative husband he is: Other husbands just moan and groan—but look what a jewel I've got! This approach is bound to succeed. However, if she can't do this, her Parent will probably retort, "When was the last time you cleaned the garage?" or her Child will send her out on the town to run up the charge accounts. This transaction illustrates that even though the Parent and Child are involved, the outcome can be amiable and advance a good marriage *if* the Adult is in charge.

The Adult has a choice as to how it will respond to a stimulus in a complementary way that will protect both the relationship and the individuals in the relationship. This sometimes takes some very rapid (intuitive) computing:

The scene is a cocktail party. The transaction is initiated by a man who (Child) pinches a woman's bottom. She responds (Adult): "My mother always told me to turn the other cheek." Why is this response identified as Adult?

She could have responded Parent: "You dirty old man!" or even slapped him.

Had she responded Child, she may have cried, become embarrassed, angry, shaky, or seductive.

Her response was Adult, however, in that she got a lot of information across in her one response.

1. I had a mother *who always told me*—so you watch out!
2. *Turn the other cheek*—I know the Bible, too, so you see I'm not that kind of girl.
3. The humor of the play on words told him, "My Child is getting a laugh, and you're OK, and I can take a joke."
4. Transaction completed!

The person who always comes out "smelling like a rose" does not do so accidentally. He has a high-speed Adult. As handy as this is in social situations, as above, it is not as critical there as in the home. You can walk away from a cocktail party. Walking away from home is something else.

The question arises: How can the Adult work better and faster? When someone knocks on the front door of life, who is going to get there first—the Parent, the Adult, or the Child?

How to Stay in the Adult

The Adult develops later than the Parent and Child and seems to have a difficult time catching up throughout life. The Parent and Child occupy primary circuits, which tend to come on automatically in response to stimuli. The first way, therefore, to build the strength of the Adult is to become sensitive to Parent and Child signals. Aroused feelings are a clue that the Child has been hooked. To know one's own Child, to be sensitive to one's own NOT OK feelings, is the first requirement for Adult data processing. Being aware that, "That is my NOT OK Child" makes it possible to keep from externalizing the feelings in actions. Processing this data takes a moment. Counting to ten is a useful way to delay the automatic response in order that the Adult maintain control of the transaction. "When in doubt, leave it out" is a good practice for curtailing archaic, or destructive, Child reactions. Aristotle claimed that the real show of power is in restraint. The strength of the Adult shows first also in restraint—in restraining the automatic, archaic responses of Parent and Child, while waiting for the Adult to compute appropriate responses.

Parent signals can be monitored in the same way. It is helpful to program into the computer certain Adult questions to apply to Parent data: Is it true? Does it apply? Is it appropriate? Where did I get that idea? What is the evidence?

The more one knows of the content of Parent and Child, the more easily one can separate Parent and Child from the Adult. In England psychotherapy is called "sorting yourself out." This is precisely the process required for developing the Adult. The more sensitive one is to one's own Parent and Child, the more separated, autonomous, and strong becomes the Adult.

One way to practice identifying the Parent and Child is to monitor the internal dialogue. This is relatively simple, inasmuch as there are no external demands for response, and one has time to examine the data. When one feels badly, gloomy, regretful, depressed, one can ask the question, "Why is my Parent beating on my Child?" Internal, accusatory dialogues are commonplace. Bertrand Russell wrote about Alfred North Whitehead: "Like other men who lead extremely disciplined lives, he was liable to distressing soliloquies and when he thought he was alone, he would mutter abuse of himself for his supposed shortcomings.*

When one is able to say, "That is my Parent," or "That is my Child," one says it with the Adult, so by the very process of questioning one has shifted to the Adult. One is able to feel immediate relief in a stressful situation simply by asking, "Who's coming on?"

As one becomes sensitive to one's own Child, one begins to become sensitive to the Child in others. No man loves the man he fears. We fear the Parent in others; their Child we can love. One helpful practice in a difficult transaction is to *see* the little boy, or the little girl, in another person, and talk to that little boy or girl, not in a condescending way but in a loving, protective way. When in doubt, stroke. When one is responding to another's Child, one is not afraid of the other's Parent.

An example of "talking to the little boy" appears in Adele Rogers St. Johns, *Tell No Man,* wherein Hank Gavin says:

> "I—I had a sort of sight of her *through* what she was now. I'd
> had this happen a couple of times on big deals with men, heads

---

*Russell, *The Autobiography of Bertrand Russell.*

of companies—I got a sight of them as though I were seeing *through*—and it was sometimes a kind of strange, wistful, desperate fellow—like the kid he'd been when he went fishing with angleworms. That may sound far out, but it had happened a couple of times, and I'd made the pitch to—to *that* fellow and it worked."*

That fellow was the Child.

Another way to strengthen the Adult is to take the time to make some big decisions about basic values, which will make a lot of smaller decisions unnecessary. These big decisions can always be re-examined, but the time it takes to make them does not have to be spent on every incident in which basic values apply. These big decisions form an ethical basis for the moment-to-moment questions of what to do.

Conscious effort is required to make these big decisions. You can't teach navigation in the middle of a storm. Likewise, you can't build a system of values in the split second between your son's statement "Johnny punched me in the nose," and your response. You can't carry through a constructive transaction with the Adult in charge if basic values and priorities have received no thought beforehand.

If you own a cruiser, you become an expert navigator because you have acquainted yourself with the consequences of being a poor one. You don't wait until the storm hits to figure out how to work the radio. If you have a marriage, you become an expert partner because you have acquainted yourself with the consequences of being a poor one. You work out a value system to underlie your marriage, which then serves you when the going gets rough. Then the Adult is prepared to take over transactions with a question such as, "What's important here?"

The Adult, functioning as a probability estimator, can work out a system of values that encompasses not only the marriage rela-

*A. Rogers St. Johns, *Tell No Man* (New York: Doubleday, 1966).

tionship but all relationships. Unlike the Child, it can estimate consequences and postpone gratifications. It can establish new values based on a more thorough examination of the historical, philosophical, and religious foundations for values. Unlike the Parent, it is concerned more with the preservation of the individual than with the preservation of the institution. The Adult can consciously commit itself to the position that to be loving is important. The Adult can see more than a parental mandate in the idea "it is more blessed to give than to receive."

The kind of giving which is Adult is reflected upon by Erich Fromm:

> The most widespread misunderstanding is that which assumes that giving is "giving up" something, being deprived of, sacrificing. People whose main orientation is a nonproductive one feel giving as an impoverishment . . . just because it is painful to give, one *should* [Parent] give; the virtue of giving, to them, lies in the very act of acceptance of sacrifice. . . .
>
> For the productive character [Adult] giving has an entirely different meaning. Giving is the highest expression of potency. In the very act of giving I experience my strength, my wealth, my power. This experience of heightened vitality and potency fills me with joy. I experience myself as overflowing, spending, alive, hence as joyous. Giving is more joyous than receiving, not because it is a deprivation, but because in the act of giving lies the expression of my aliveness [OK].*

This kind of giving can be a chosen way of life. This choice can underlie all decisions as the Adult asks: What is important here? Am I being loving? Once such value decisions are made one can constructively intercept "Where did you hide the can opener?" and proceed with a day-to-day strengthening of the I'M OK—YOU'RE OK position.

In summary, a strong Adult is built in the following ways:

*E. Fromm, *The Art of Loving* (New York: Harper, 1966).

1. Learn to recognize your Child, its vulnerabilities, its fears, its principal methods of expressing these feelings.

2. Learn to recognize your Parent, its admonitions, injunctions, fixed positions, and principal ways of expressing these admonitions, injunctions, and positions.

3. Be sensitive to the Child in others, talk to that Child, stroke that Child, protect that Child, and appreciate its need for creative expression as well as the NOT OK burden it carries about.

4. Count to ten, if necessary, in order to give the Adult time to process the data coming into the computer, to sort out Parent and Child from reality.

5. When in doubt, leave it out. You can't be attacked for what you didn't say.

6. Work out a system of values. You can't make decisions without an ethical framework. How the Adult works out a value system is examined in detail in Chapter 12, "P-A-C and Moral Values."

6

How We Differ

The tools of the mind become burdens when the environment which made them necessary no longer exists.

—Henri Bergson

All people are structurally alike in that they all have a Parent, an Adult, and a Child.

They differ in two ways: in the content of Parent, Adult, and Child, which is unique to each person, being recordings of those experiences unique to each; and in the working arrangement, or the functioning, of Parent, Adult, and Child.

This chapter is devoted to an examination of these *functional differences*. There are two kinds of functional problems: *contamination* and *exclusion*. *

Contamination

At the end of an initial hour in which I had explained P-A-C to a sixteen-year-old girl who was withdrawn, uncommunicative, cultur-

---

*E. Berne, *Transactional Analysis in Psychotherapy* (New York: Grove Press, 1961).

ally deprived, a school dropout, and referred by the Welfare Department, I asked, "Can you tell me what P-A-C means to you now?"

After a long silence she said, "It means that we are all made up of three parts and we'd better keep them separated or we're in trouble."

The trouble when they are not separated is called *contamination of the Adult.*

Ideally (Figure 30) the P-A-C circles are separate. In many people, however, the circles overlap. The (a) overlap in the figure is contamination of the Adult by dated, unexamined Parent data which is externalized as true. This is called *prejudice.* Thus, beliefs such as "white skins are better than black skins," "right-handedness is better than left-handedness," and "cops are bad" are externalized in transactions on the basis of prejudgment, before reality data (Adult) is applied to them. Prejudice develops in early childhood when the door of inquiry is shut on certain subjects by the security-giving parents. The little person dares not open it for fear of parental rebuke.

We all know how difficult it is to *reason* with a prejudiced person. With some people one can present a logical and evidential case regarding racial issues or left-handedness or any other subject that the person holds a prejudice about; yet, the Parent in these people steadfastly dominates a portion of the Adult, and they will surround their prejudicial cases with all kinds of irrelevant arguments to support their position. As illogical as their position may seem, the rigidity of their position is in its *safety*. As illustrated in Chapter 2, it is safer for a little child to believe a lie than to believe his own eyes and ears. Therefore, one cannot eliminate prejudice by an Adult discourse on the subject of the prejudice. The only ways to eliminate prejudice are to uncover the fact that it is no longer dangerous to disagree with one's parents and to update the Parent with data from today's reality. Thus, treatment can be seen as separating Parent and Adult and restoring the boundary between them.

The (b) overlap in Figure 30 is contamination of the Adult by the Child in the form of feelings or archaic experiences which are inappropriately externalized in the present. Two of the most common symptoms of this kind of contamination are *delusions* and *hallucinations*. A delusion is grounded in fear. A patient who said to me, "The world is hideous," was describing how the world seemed to him as a small child. A little person who was in constant fear of brutality at the hands of angry, unpredictable parents can, as a grownup, under stress, be flooded by the same fear to the extent that he can fabricate "logical" supporting data. He may believe that the door-to-door salesman coming down the street is really coming to kill him. If confronted with the fact that it is only a salesman, this person may support his fear by a statement such as "I knew it the minute I saw him. It's him! He's wanted by the FBI. I saw his picture in the post office. That's why he's coming to get me." As in the case of prejudice, this delusion cannot be eliminated by a simple

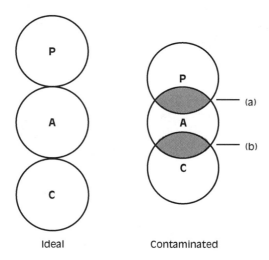

Ideal   Contaminated

Fig. 30. Contamination.
(a) Prejudice
(b) Delusions

statement of the truth that this is, in fact, a salesman. It can only be eliminated by uncovering the truth that the original threat to the Child no longer exists externally. Only as the Adult is decontaminated is it able to compute reality data.

Hallucinations are another type of contamination of the Adult by the Child. A hallucination is a phenomenon produced by extreme stress, wherein what was once experienced externally—derogation, rejection, criticism—is again experienced externally, even though "no one is there." A recorded experience "comes on for real" and the person "hears" voices that existed in a past reality. If you ask him what the voices say, he characteristically will describe the content as words of criticism, threat, or violence. The more bizarre the hallucination the more bizarre was life for him as a child. Bizarre hallucinations are not hard to understand when we consider the actual types of abuse, verbal and physical, to which some children are subjected.

Exclusion

In addition to contamination there is another functional disorder that explains how we differ: *exclusion.*

> Exclusion is manifested by a stereotyped, predictable attitude which is steadfastly maintained as long as possible in the face of any threatening situation. The constant Parent, the constant Adult, and the constant Child all result primarily from defensive exclusion of the two complementary aspects in each case.*

This is a situation in which an Excluding Parent can "block out" the Child or an Excluding Child can "block out" the Parent.

*Berne, *Transactional Analysis in Psychotherapy.*

The Person Who Cannot Play

Typical of the Parent-Contaminated Adult with a Blocked-Out Child (Figure 31) is the man who is duty-dominated, always working late at the office, all business, impatient with family members who want to plan a skiing trip or a picnic in the woods. It is as if, at some point in his childhood, he was so utterly quashed by serious, stern, duty-bound parents that he found the only safe way to proceed through life was to turn his Child off completely, or to block it out. He had found, through experience, that every time he let it out there was trouble: "Go to your room"; "Children should be seen and not heard"; "How many times must I tell you . . ."; *"Grow up!"* If this little person also was rewarded for perfect conformity, diligent effort, compliance, and doing exactly as he was told, the path of wisdom appeared to be total conformity to the Parent and total blocking out of childlike impulses.

This type of person has very little happiness recorded in his Child. He probably never will be able to let his happy Child out,

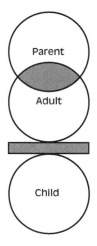

Fig. 31. Parent-contaminated Adult
with a blocked-out Child.

because he has very little happy Child. However, he can be helped to appreciate that his position is not fair to his own family and to his own children and that, indeed, his marriage may break up if he persists in trying to block out the Child in his wife and children. He can, through purposeful effort by his Adult, take a trip with his family, cut down his office hours, listen lovingly (an Adult discipline) to his children's fantasies, and participate in their life. He can, with his Adult, establish a value of being loving or of preserving his family. He cannot change the content of his Parent or create happy Child recordings which are not within him, but he can achieve the insight which makes it possible to build a satisfactory life in the present.

The Person Without a Conscience

A more serious difficulty, particularly to society, is presented by the Child-Contaminated Adult with a Blocked-Out Parent (Figure 32). This condition develops in the person whose real parents, or

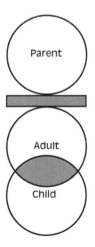

Fig. 32. Child-contaminated Adult with a blocked-out Parent.

those who fulfilled the parental role, were so brutal and terrifying or, in the other extreme, so doltishly indulgent that the only way to preserve life was to "shut them off" or block them out. This is typical of the psychopath, the person who at some point early in life gives up the first position, I'M NOT OK—YOU'RE OK, and assumes a new one, I'M OK—YOU'RE NOT OK. This little person concludes correctly that his parents are in fact NOT OK. They are so NOT OK that he excludes them entirely. In the extreme he may do this by homicide. If not this, he at least excludes them psychologically, so that, in a sense, he does not have a Parent. He excludes the painful Parent, but he also excludes what little "good" there is in the Parent. Such a person does not have available to his current transactions any tapes which supply data having to do with social control, appropriate "shoulds" and "should nots," cultural norms, or what, in one sense, may be referred to as conscience. His behavior is dominated by his Child which, through the contaminated Adult, manipulates other people to his own ends. His Adult is able to estimate consequences, but the consequences he is concerned with have to do with whether or not he will be caught and seldom contain elements of concern for others. Although there may be exceptions, the general rule is that we do not learn to be loving if we have never been loved. If the first five years of life consist totally of a critical struggle for physical and psychological survival, this struggle is likely to persist throughout life.

One way to determine whether or not a person has a Parent is to determine the existence of feelings of shame, remorse, embarrassment, or guilt. These feelings, which exist in the Child, turn on as the Parent "beats on the Child." If these feelings do not exist, it is probable the Parent has been blocked out. It is a safe assumption that if a man who has been arrested for child-molestation does not express any feelings of remorse or guilt—apart from the fact that he was caught—he does not have a functioning Parent. This has prognostic implications for rehabilitation. The treatment of such a person is difficult. One cannot evoke a Parent where one does not exist. A number of experiments have been conducted with monkeys who were raised not by their real mothers but by surrogate mothers in the form of wire dummies covered with terry cloth. During infancy the little monkeys formed strong attachment to these terry-cloth surrogates. However, when these monkeys, reared by the terry-cloth mothers, reached maturity, their capacity for reproduction and rearing their own young was minimal.* They were deficient in experience about mothering, often thought to be instinctual. Mothering was not recorded in the Parent, so nothing replayed.

The prognosis for the person with a blocked-out Parent is not quite so dire in that, unlike the monkey, he has a 12-billion-cell computer with which to assess reality and construct answers even if none were recorded early. A criminal psychopath *can* understand his (P)-A-C to the extent that his Adult can direct his future activities in such a way that his pattern of crime, arrest, and conviction can change. He may never have an operational Parent to back up his Adult, but his Adult can become strong enough to carry him through a successful life wherein he gains the approval and even esteem of others. It is on this possibility that rehabilitative efforts in the field of corrections must be based.

The Decommissioned Adult

The person who has a Blocked-Out Adult (Figure 33) is psychotic. His Adult is not functioning, and therefore he is out of touch with reality. His Parent and Child come on straight, frequently in a jumbled mixture of archaic data, a jumbled replay of early experiences that do not make sense now because they did not make sense when they were recorded. This was observable in a female hospital patient whose singing of tent-meeting hymns (Parent) was inter-

*H. F. Harlow, "The Heterosexual Affectional System in Monkeys," *American Psychologist* 17 (1962): 1–9.

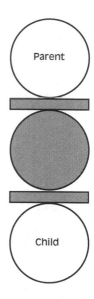

Fig. 33. The blocked-out, or decommissioned,
Adult (psychosis).

spersed with obscenities relating to body functions (Child). The content was bizarre, but seemed to replay an old Parent-Child conflict between good and bad, should and should not, salvation and damnation. The content of these verbal productions quickly revealed a great deal about her Parent and Child. The fact that her Adult was gone indicated the severity of the conflict. "The struggle is too hard; I will not try." This is not to say that there was any comfort in her giving up. She was at the mercy of the same terrifying feelings which existed in her as a child.

The first step in treating a psychotic patient is to reduce these feelings of terror. Basic to the recovery of the patient is that from his first encounter with the therapist he sense the demonstrated position I'M OK—YOU'RE OK. In 1963 my associate, Gordon Haiberg, reported the effect of this stated position on psychotic patients he was then treating at Stockton State Hospital:

Treatment begins with the first exchange of glances between the therapist and patient, when the therapist enters with the basic position I'M OK—YOU'RE OK. Psychotics are yearning to establish a more meaningful relationship with people. . . . When these usually very perceptive individuals are confronted by an individual who assumes the position I'M OK—YOU'RE OK, this is a new and intriguing experience for them. The Adult is "hooked" by this human assumption and begins asking, "How come you're on the outside and I'm on the inside?" The question is not answered immediately but at the intuitively correct time after the therapist has had the opportunity to establish what is the most destructive maneuver or game the person is using. The patient is simply and directly informed, for example: "You scare the hell out of people." . . . He is frankly told he is OK, he is important because of the uniqueness of his being a human being, and he starts to gain hope. When the Adult can begin processing data, listening, learning, and helping in decisions, the innate regenerative powers begin operating on their own and the healing process begins.*

My ongoing hospital groups, which at any given time contain from ten to twenty patients, frequently include psychotic patients. In the setting of a group, wherein I'M OK—YOU'RE OK is the stated contract, the psychotic patient feels supported, stroked, and reassured, and the stage is set for the return of the Adult.

In one hospital group one woman got up repeatedly throughout the hour to straighten her skirt and pull it down below her knees. Though this activity could not go unnoticed, no one made anything of it, no one looked disapproving; the group simply went on with its discussion. The hour included thirty minutes of the teaching of P-A-C, followed by discussion. At the end of the hour the man next to the woman said to her, "You know, I kept track, and you got up to straighten your skirt nineteen times."

*G. Hailberg, "Transactional Analysis with State Hospital Psychotics," *Transactional Analysis Bulletin,* Vol. 2, No. 8 (October 1963).

With some surprise and delight in discovery she said, "I did?" She was able to make this Adult response without suspicion of threat. The emergence of the Adult, however tenuous, is the beginning of the re-establishment of contact with reality, and the stage is set for the learning of P-A-C, through which further discovery can be made by the patient about his own behavior.

The Constant Bore

There is a type of bland individual whose recordings in the Parent and Child are so insipid that he intrinsically lacks the raw materials for a colorful personality. This most often expresses itself clinically in the person who has a vague depression (happiness is for other people) or simply is bored with life. His parents were dull, inarticulate, and ambivalent. There was rarely punishment and rarely reward. There was little enthusiasm about anything. As a little person he was not exposed to the excitement of an external world, he seldom had relationships with other children, and though he was a "good" boy, in that he didn't cause particular trouble, he was not a child that anyone would notice. His Adult correctly perceived reality, but the reality itself was dull. He may well have grown up to have an emancipated Adult but one that sees no positive value in reaching out to other people. (This kind of value is generally first seen in the Parent, if indeed it is a value of the Parent.) His personality is much like a computer. While others enjoy themselves at a party he leafs through a magazine in a corner doing the only thing he knows how to do—sift data. If he ultimately comes to treatment his question may be something like, "Isn't there really more than this?" Though he is no problem to society, he is a problem to himself. His reality is limited in the same way that his reality was limited early in life.

In a sense Alfred North Whitehead's statement applies here: "Moral education is impossible apart from the habitual vision of greatness." If "moral" is thought of in terms of a value system and if "not being a bore" or "being interesting to other people" or "being creative and productive" are seen as positive values, it is clear that a person whose earliest impressions of life were dull will, unless some spectacular relationships appear, be dull himself.

It is true there are other reasons for boredom and for being a bore. A little person who begins life with vast curiosity about why smoke goes up, why rain comes down, who made God, and who made me and who consistently is given pat answers that only add to his confusion and turn out later to be wrong, eventually quits asking, quits exploring, quits being interested, and begins being bored. His computer begins shutting down on vast areas of interest because the answers to his questions have only added to his confusion. This type of boredom is often evidenced among high school and college students in the setting of the church in which they grew up. Their boredom grows from the simplistic answers they frequently are given in the quest for truth, the inhibitions which are imposed on their following truth (evidential, observable data) where it takes them, and the edict that they must choose between truth and faith, as if these were mutually exclusive. Not all clergymen give simplistic answers; yet unexamined dogma still persists as the rule in many religious communities. This will be further elaborated in Chapter 12, "P-A-C and Moral Values."

Differences in the Content of P-A-C

In the foregoing we have seen how people differ by virtue of different structural arrangements of Parent, Adult, and Child. Most of these differences have appeared as clinical problems. We differ also in "healthy" ways. The definition of health is an emancipated Adult that is consistently in charge of every transaction. This means that in every transaction the Adult takes data from the Parent, from the Child, and from reality and comes up with a decision as to what to do. The richer the fund of data to be drawn from, the more possibilities for fulfillment exist. The little child whose early experiences included unhampered exploration of the pots and pans, the mud along with the daisies, pets, friends, trips to the farm, evenings of storytelling, tradition surrounding holidays, toys to manipulate, records to listen to, and open and friendly conversations with unhurried parents will have a far richer fund of data in his Parent and a great many more positive feelings in his Child than the little person who is isolated and overprotected. The little person who early tests ways of overcoming the NOT OK position strengthens his Adult and is encouraged to further exploration and mastery. He then becomes a "bright child," bringing to himself the praise and self-confidence which inspire him to be brighter. The fact that he brings credit to the family does not minimize the fact that he brings credit to himself, unless, in the process, his parents apply undue pressure and put demands upon him that are truly not in his self-interest.

Through understanding our own P-A-C, we can come to understand not only what is in the Parent and the Child, but also what *is not* there. If the girl who bemoans the fact that "I am plain and dull . . . that's just the way I am" can appreciate the deficiency in her Parent and Child because her introduction to life was plain and dull, she can then be free, with her Adult, to reach out to reality to discover what is not plain and dull. It may take her a while to catch up, and she won't turn into the life of the party overnight, but she can certainly be helped to see that *she has a choice.* It has been said that blaming your faults on your nature does not change the nature of your faults. Thus, "I am like that" does not help anything. "I can be different" does.

7

How We Use Time

Time is what we want most, but what alas! we use worst.

—William Penn

One of the most dramatic scientific adventures of this century is the exploration of space. We are not content to understand that it is infinite. We want landmarks, so to speak, platforms for our satellites, or mathematical slots into which we can aim our space vehicles. We want to comprehend space; to define it; in a sense, to use it.

The other great cosmic reality is time. We may speculate about either end of our earthly existence. We may trust in immortality in the face of incomprehensible death; but, as in our efforts to define space, we must in our definition of time start where we are. All we can know is that man's average ration of time is three score and ten years. What we do with our known allotment is what concerns us. Of most immediate concern is what we do with the smaller blocks of time within our grasp: the next week, the next day, the next hour, this very hour.

We all share with Disraeli a common concern that "life is too short to be little." Yet our greatest frustration is that so much of life is just that. Perhaps more significant and dramatic than space exploration is an investigation of our use of time. "What folly," said John Howe, "to dread the thought of throwing away life at once, and yet have no regard to throwing it away by parcels and piecemeal." As with space, we are not content to comprehend time only as infinite. For many people the pressing question is "How am I going to get through the next hour?" The more structured time is, the less difficult is this problem. Very busy people with many external demands do not have time on their hands. The "next hour" is very well programed. This programing, or structuring, is what people try to achieve, and when they are unable to do it themselves, they look to others to structure time for them. "Tell me what to do." "What shall I do next?" "What we need is leadership."

Structure hunger is an outgrowth of recognition hunger, which grew from the initial stroking hunger. The small child has not the necessary comprehension of time to structure it but simply sets about doing things which feel good, moment to moment. As he gets a little older he learns to postpone gratifications for greater rewards: "I can go outside and make mudpies with Susie now, but if I wait twenty more minutes and keep my nice dress on, I can go to the shopping center with Daddy." This is basically a problem in structuring time. Which alternative will be more fun? Which will bring a greater reward? As we grow older we have more and more choices. However, the NOT OK position keeps us from exercising these choices as freely as we might think we do.

In our observation of transactions between people, we have been able to establish six types of experience, which are inclusive of all transactions.

They are withdrawal, rituals, activities, pastimes, games, and intimacy.

Withdrawal, although it is not a transaction with another person, can take place, nonetheless, in a social setting. A man, having lunch with a group of boring associates more concerned about their own stroking than his, may withdraw into the fantasy of the night before, when the stroking was good. His body is still at the lunch table, but "he" isn't. Schoolrooms on a nice spring day are filled with bodies whose "occupants" are down at the swimming hole, shooting into space on a blazing rocket, or recalling how nice it was kissing under the wisteria. Whenever people withdraw in such fashion it is always certain that the withdrawal keeps them apart from those they are with bodily. This is fairly harmless unless it happens all the time, or unless your wife is talking to you.

A ritual is a socially programed use of time where everybody agrees to do the same thing. It is safe, there is no commitment to or involvement with another person, the outcome is predictable, and it can be pleasant insofar as you are "in step" or doing the right thing. There are worship rituals, greeting rituals, cocktail party rituals, bedroom rituals. The ritual is designed to get a group of people through the hour without having to get close to anyone. They may, but they don't have to. It is more comfortable to go to a High Church Mass than to attend a revival service where one may be asked, "Are you saved, brother?" Sexual relations are less awkward in the dark for people for whom physical intimacy has no involvement at the level of personality. There is less chance for involvement in throwing a cocktail party than in having a dinner for six. There is little commitment, therefore little fulfillment. Rituals, like withdrawal, can keep us apart.

An activity, according to Berne, is a "common, convenient, comfortable, and utilitarian method of structuring time by a project designed to deal with the material of external reality."* Common activities are keeping business appointments, doing the dishes, building a house, writing a book, shoveling snow, and studying for mid-terms. These activities, in that they are productive or creative, may be highly satisfying in and of themselves, or they may lead to satisfactions in the future in the nature of stroking for a job well

---

*E. Berne, *Transactional Analysis in Psychotherapy* (New York: Grove Press, 1961), p. 85.

done. But during the time of the activity, there is no need for intimate involvement with another person. There may be, but there does not have to be. Some people use their work to avoid intimacy, working nights at the office instead of coming home, devoting their lives to making a million instead of making friends. Activities, like withdrawal and rituals, can keep us apart.

Pastimes are a way of passing time. Berne defines a pastime as

> . . . an engagement in which the transactions are straightforward. . . . With happy or well-organized people whose capacity for enjoyment is unimpaired, a social pastime may be indulged in for its own sake and bring its own satisfactions. With others, particularly neurotics, it is just what the name implies, a way of passing (i.e. structuring) the time: until one gets to know people better, until this hour has been sweated out, and on a larger scale, until bed-time, until vacation time, until school starts, until the cure is forthcoming, until some form of charism, rescue, or death arrives. Existentially a pastime is a way of warding off guilt, despair, or intimacy, a device provided by nature or culture to ease the quiet desperation. More optimistically, at best it is something enjoyed for its own sake and at least it serves as a means of getting acquainted in the hope of achieving the longed-for crasis with another human being. In any case, each participant uses it in an opportunistic way to get whatever primary and secondary gains he can from it.*

People who cannot engage in pastimes at will are not socially facile. Pastimes can be thought of as being a type of social probing where one seeks information about new acquaintances in an unthreatening, noncommittal way. Berne's observation is that "pastimes form the basis for the selection of acquaintances and may lead to friendship" and further that they have as an advantage the "confirmation of role and the stabilizing of position." Berne has given some delightful and disarming names to cer-

*Berne, *Transactional Analysis in Psychotherapy*, p. 98.

tain of these pastimes, which can be recognized at cocktail parties, women's luncheons, family reunions, and the Kiwanis Club as: variations of "Small Talk," such as "General Motors" (comparing cars) and "Who Won" (both "man talk"); "Grocery," "Kitchen," and "Wardrobe" (all "lady talk"); "How To" (go about doing something); "How Much" (does it cost?); "Ever Been" (to some nostalgic place); "Do You Know" (So-and-So); "What Became Of" (Good Old Joe); "Morning After" (what a hangover!); and "Martini" (I know a better way).*

Pastimes may be played by the Parent, Adult, or Child. A Parent-Parent pastime was initiated by the following transaction:

MAUDE: You mean you do upholstery?

BESS: Only when necessary.

This led to a discussion of the high price of having it done, how shoddy work is these days, and the sale at Macy's.

One Child-Child pastime is the sharing of impossible alternatives symbolic of the damned-if-you-do and damned-if-you-don't situation of the little child. Anxiety may be relieved by this pastime, not because the problem is solved, but because it is handed to someone else—"Here, *you* struggle with this for a while!" The following questions were overheard in an exchange between two five-year-olds: Would you rather eat a hill full of ants or drink a pail of boiling medicine? Would you rather be chased by a wild bull or wear your shoes on the wrong feet all day? Would you rather sit on a hot stove or go through the washing machine fifty times? Would you rather be stung by a thousand wasps or sleep in the pigpen? *Answer one or the other!* You *have* to answer one or the other. Grownup versions may be more sophisticated, as, Are you a Democrat or a Republican?

The Adult may play pastimes about such subjects as the

*Berne, *Transactional Analysis in Psychotherapy,* p. 99.

weather in order to keep a relationship going until something interesting or stroke-producing appears:

MR. A: Looks like a storm coming up.

MR. B: Those clouds really look black.

MR. A: Reminds me of the time I was flying my plane and ran into a squall over San Francisco Bay.

MR. B: Oh, you fly?

As useful as pastimes may be in certain social situations, it is evident that relationships that do not progress beyond them die or, at best, exist in quiet desperation and growing boredom. Pastimes, like withdrawal, rituals, and activities, can keep people apart.

Games are such significant transactional phenomena that Berne has devoted a whole book to them, his best-selling *Games People Play*. Most games cause trouble. They are the relationship wreckers and the misery producers, and in understanding them lies the answer to "why does this always happen to me?" The word "game" should not be misleading, explains Berne. It does not necessarily imply fun or even enjoyment. For a full understanding of games, his book is recommended. However, the following is a brief definition, which will serve the purposes of this guide to Transactional Analysis.

A game is an ongoing series of complementary ulterior transactions progressing to a well-defined, predictable outcome. Descriptively it is a recurring set of transactions, often repetitious, superficially plausible, with a concealed motivation; or, more colloquially, a series of moves with a snare, or "gimmick." Games are clearly differentiated from procedures, rituals, and pastimes by two chief characteristics: (1) their ulterior quality and (2) the payoff. Procedures may be successful, rituals effective, and pastimes profitable, but all of them are by definition candid; they may involve contest, but not conflict, and the end-ing may be sensational, but it is not dramatic. Every game, on the other hand, is basically dishonest, and the outcome has a dramatic, as distinct from merely exciting, quality.*

As pointed out in Chapter 3, all games have their origin in the simple childhood game of "Mine Is Better Than Yours," easily observable in any group of four-year-olds. It was then, as it is now, designed to bring a little momentary relief from the burden of the NOT OK position. As in the more sophisticated grownup versions, it is ulterior in that it does not express what is really felt. When the little person says, "Mine is better than yours," he is really feeling, "I'm not as good as you." It is an offensive defense. It is protective in that it seeks to maintain homeostasis. It also has a pay-off, as do games grownups play. When "Mine Is Better Than Yours" is pushed far enough, the game ends with a hard shove, a slapped face, or devastating evidence of some sort that "It is not: *mine's better.*" This then puts the little person back in his place, it has been proved again that I'M NOT OK, and in the maintenance of this fixed position there is a certain miserable security.

This is the essence of all games. Games are a way of using time for people who cannot bear the stroking starvation of withdrawal and yet whose NOT OK position makes the ultimate form of relatedness, intimacy, impossible. Though there is misery, there is something. As the comedian said, "It's better to have halitosis than no breath at all." It is better to be roughed up playing games than to have no relationship at all. "The developing [child] is more likely to survive in the warmth of wrath and to suffer blight in the chill of indifference," wrote Dr. Richard Galdston, of abused children.[†]

Thus, games provide benefits to all the players. They protect the integrity of the position without the threat of uncovering the position.

*E. Berne, *Games People Play* (New York: Grove Press, 1964), p. 48.
†R. Galdston, M.D., "Observations of Children Who Have Been Physically Abused and Their Parents," *American Journal of Psychiatry*, Vol. 122, No. 4 (October 1965).

To further clarify the nature of games we shall report the moves in one game, "Why Don't You, Yes But." The players are Jane, a young career woman, and her friend. (This game frequently is played in the helping situation, the clergyman's study, the psychiatrist's office, or the kitchen of a long-suffering coffee mate.)

JANE: I am so plain and dull that I never have any dates.

FRIEND: Why don't you go to a good beauty salon and get a different hairdo?

JANE: Yes, but that costs too much money.

FRIEND: Well, how about buying a magazine with some suggestions for different ways of setting it yourself?

JANE: Yes, but I tried that—and my hair is too fine. It doesn't hold a set. If I wear it in a bun, it at least looks neat.

FRIEND: How about using makeup to dramatize your features, then?

JANE: Yes, but my skin is allergic to makeup. I tried it once and my skin got rough and broke out.

FRIEND: They have lots of good new nonallergenic makeups out now. Why don't you go see a dermatologist?

JANE: Yes, but I know what he'll say. He'll say I don't eat right. I know I eat too much junk and don't have well-balanced meals. That's the way it is when you live by yourself. Oh, well, beauty is only skin deep.

FRIEND: Well, that's true. Maybe it would help if you took some Adult Education courses, like in art or current events. It helps make you a good conversationalist, you know.

JANE: Yes, but they're all at night. And after work I'm so exhausted.

FRIEND: Well, take some correspondence courses, then.

JANE: Yes, but I don't even have time to write letters to my folks. How could I ever find time for correspondence courses?

FRIEND: You could find time if it were important enough.

JANE: Yes, but that's easy for you to say. You have so much energy. I'm always dead.

FRIEND: Why don't you go to bed at night? No wonder you're tired when you sit up and watch "The Late Show" every night.

JANE: Yes, but I've got to do *something* fun. That's all there is to do when you're like me!

Here the discussion has gone full circle. Jane has systematically knocked down every one of her friend's suggestions. She begins with the complaint that she is plain and dull, then ends up begging the question with the final reason: she is plain and dull because "that is the way I am."

Her friend finally gives up in defeat and perhaps finally stops coming over, further underlining Jane's NOT OK. This "proves" to Jane that there indeed is no hope for her—she can't even keep the friends she has, and this justifies her indulging in still another game, "Ain't It Awful." The benefit to Jane is that she doesn't have to do anything about herself because she has repeated proof that nothing can be done.

"Why Don't You, Yes But" can be played by any number, according to Berne:

> One player, who is "it," presents a problem. The others start to present solutions, each beginning with "Why Don't You." To each of these the one who is "it" objects with a "Yes But." A good player can stand off the rest of the group indefinitely, until they all give up, whereupon "it" wins.

Since all the solutions, with rare exception, are rejected, it is apparent that this game must serve some ulterior purpose. The "gimmick" in "Why Don't You, Yes But" is that it is not played for its ostensible purpose (an Adult quest for information or solutions) but to reassure and gratify the Child. A bare transcript may sound Adult, but in living tissue it can be observed that the one who is "it" presents herself as a Child inadequate to meet the situation; whereupon the others become transformed into sage Parents eager to dispense their wisdom for the benefit of the helpless one. This is exactly what "it" wants, since her object is to confound these Parents one after another.*

(This is a latter-day version of "Mine's Better Than Yours," which denies the real conviction, You Are Better Than I.) As the game ends, all those who offered advice are dejected, having failed in helping "it," and "it" has proved the point that her problem really is insoluble, which makes it possible for her to indulge her Child in a new game of "Ain't It Awful." That's the way it is and that's the way I am (and therefore I don't have to do anything about it, for, as we have just seen, nothing can be done).

Berne describes about three dozen games in *Games People Play*. His game titles are colloquial, and most of them, with semantic precision, put the finger on the central characteristic of the game, as: "Ain't It Awful"; "If It Weren't for You, I Could"; "Let's You and Him Fight"; and "Now I've Got You, You Son of a Bitch." Because the titles are colloquial, they frequently bring a laugh. The fact is games are not funny. They are defenses to protect individuals from greater or lesser degrees of pain growing from the NOT OK position. The popularity of Berne's game book has given rise in many sophisticated circles to a new pastime of game calling. The concept of games can be a useful therapeutic tool when used in combination with a prior applied understanding of P-A-C; but in the absence of such insight the game concept, particularly game calling, can simply be another way to be hostile. People with an understanding of P-A-C can use an academic discussion of games by

*Berne, *Transactional Analysis,* p. 104.

applying it to themselves; but to be "called" on a game by another person, in the absence of insight or true concern, most often produces anger. It is my firm belief from long observation of this phenomenon that game analysis must always be secondary to Structural and Transactional Analysis. Knowing what game you are playing does not, *ipso facto,* make it possible for you to change. There is danger in stripping away a defense without first helping a person to understand the position—and the situation in childhood in which it was established—which has made this defense necessary. Another way of stating this is that if only one hour were available to help someone, the method of choice would be a concise teaching of the meaning of P-A-C and the phenomenon of the transaction. This procedure, I believe, holds more promise for change in short-term treatment than game analysis.

Summarily, we see games as time-structuring devices which, like withdrawal, rituals, activities, and pastimes, keep people apart. What then can we do with time in a way which does not keep us apart? George Sarton observed: "I believe one can divide men into two principal categories: those who suffer the tormenting desire for unity and those who do not. Between these two kinds an abyss— the 'unitary' is the troubled; the other is the peaceful."

For many thousands of years man's existence has been structured preponderantly by withdrawal, ritual, pastimes, activities, and games. Skepticism about this assertion could perhaps best be met by a reminder of the persistent recurrence throughout history of war, the grimmest game of all. The majority of men have helplessly accepted these patterns as human nature, the inevitable course of events, the symptoms of history repeating itself. There has been a certain peace in a resignation of this sort. But, as Sarton suggests, the truly troubled people of history have been those who have refused to resign themselves to the inevitability of apartness and who have been driven on by a tormenting desire for unity. The central dynamic of philosophy has been the impulse to connect. The hope has always been there, but it has not overcome the intrinsic fear of being close, of losing oneself in another, of partaking in the last of our structuring options, intimacy.

A relationship of intimacy between two people may be thought of as existing independent of the first five ways of time structuring: withdrawal, pastimes, activities, rituals, and games. It is based on the acceptance by both people of the I'M OK—YOU'RE OK position. It rests, literally, in an accepting love where defensive time structuring is made unnecessary. Giving and sharing are spontaneous expressions of joy rather than responses to socially programed rituals. Intimacy is a game-free relationship, since goals are not ulterior. Intimacy is made possible in a situation where the absence of fear makes possible the fullness of perception, where beauty can be seen apart from utility, where possessiveness is made unnecessary by the reality of possession.

It is a relationship in which the Adult in both persons is in charge and allows for the emergence of the Natural Child. In this regard the Child may be thought of as having two natures: the Natural Child (creative, spontaneous, curious, aware, free of fear) and the Adaptive Child (adapted to the original civilizing demands of the Parent). The emancipation of the Adult can enable the Natural Child to emerge once more. The Adult can identify the demands of the Parent for what they are—archaic—and give permission to the Natural Child to emerge again, unafraid of the early civilizing process, which turned off not only his aggressive antisocial behavior but his joy and creativity as well. This is the truth that makes him free—free to be aware again and free to hear and feel and see in his own way. This is a part of the phenomenon of intimacy. Thus the gift of a handful of primroses may more readily be a spontaneous expression of love and joy than the expensive perfume from I. Magnin on the socially important anniversary date. The forgotten anniversary date is not a catastrophe for the intimate husband and wife, but it very often is for those whose relationship exists by virtue of ritual.

The question is frequently asked: Are withdrawal, pastimes,

rituals, activities, and games always bad in a relationship? It is safe to say that games nearly always are destructive, inasmuch as their dynamic is ulterior, and the ulterior quality is the antithesis of intimacy. The first four are not necessarily destructive unless they become a predominant form of time structuring. Withdrawal can be a relaxed, restorative form of solitary contemplation. Pastimes can be a pleasant way of idling the social motor. Rituals can be fun—birthday parties, holiday tradition, running to meet Dad when he's home from work—in that they repeat again and again joyous moments which can be anticipated, counted on, and remembered. Activities, which include work, not only are necessities of life but are rewarding in and of themselves, as they allow for mastery, excellence, and craftsmanship and the expression of a great variety of skills and talents. However, if there is discomfort in a relationship between two people when these modes of time structuring cease, it is safe to say there is little intimacy. Some couples program their entire time together with frantic activity. The activity itself is not destructive unless the compulsion to keep busy is one and the same as the compulsion to keep apart.

The question now arises: If we strip ourselves of the first five ways of time structuring, do we automatically have intimacy? Or do we have nothing? There seems to be no simple way to define intimacy, yet it is possible to point to those conditions which are most favorable for its appearance: the absence of games, the emancipation of the Adult, and the commitment to the position I'M OK—YOU'RE OK. It is through the emancipated Adult that we can reach out to the vast areas of knowledge about our universe and about each other, explore the depths of philosophy and religion, perceive what is new, unrefracted by the old, and perhaps find answers, one at a time, to the great perplexity, "What's the good of it all?" An elaboration of this idea will follow in Chapter 12.

7/20 Ch 8+9

8

P-A-C and Marriage

We promise according to our hopes and perform according to our fears.

—Francois, Duc de la Rochefoucauld

A friend of mine tells the following story about something that happened when he was a little boy. At the end of a meal his mother announced to the brothers and sisters, who numbered five, that dessert would be the remainder of a batch of her special homemade oatmeal cookies, whereupon she procured the cookie jar and set it on the table. There followed a noisy scramble by the children to get into the jar, with the littlest brother, age four, last, as usual. When he got to the jar he found only one cookie left, and it had a piece missing, whereupon he grabbed it and tearfully threw it to the floor in a rage of despair, crying, "My cookie is all broke!"

It is the nature of the Child to mistake disappointment for disaster, to destroy the whole cookie because a piece is missing or because it isn't as big, as perfect, or as tasty as someone else's cookie. In his family the anecdote lived as a standard retort to further complaints, "What's the matter, your cookie broke?"

This is what happens when marriages break. The Child takes

over in one or both partners, and the whole marriage is shattered when imperfections begin to appear.

Marriage is the most complicated of all human relationships. Few alliances can produce such extremes of emotion or can so quickly travel from professions of the utmost bliss to that cold, terminal legal write-off, mental cruelty. When one stops to consider the massive content of archaic data which each partner brings to the marriage through the continuing contribution of his Parent and Child, one can readily see the necessity of an emancipated Adult in each to make this relationship work. Yet the average marriage contract is made by the Child, which understands love as something you feel and not something you do, and which sees happiness as something you pursue rather than a by-product of working toward the happiness of someone other than yourself. Fortunate and rare are the young partners whose Parent contains the impressions of what a good marriage is. Many persons have never seen one. So they borrow a concept of marriage from the highly romanticized fiction they read, wherein husband has a nice job as a junior executive in a large advertising company and comes home every night with a bouquet of roses to a slender, radiant wife awaiting him in the fifty-thousand-dollar home with Armstrong floors and sparkling windows, in which the candles are lit and the stereo is playing music to make love by. When the illusion begins to shatter, when the carpets are worn hand-me-downs from the in-laws and the stereo won't work and husband loses his job and stops saying "I love you," the Child comes on with the "broken cookie" tape and the whole show ends with everything in little pieces. What is borrowed is the illusion and what is blue is the Child. Archaic feelings of NOT OK contaminate the Adult in each partner, and, having no where else to turn, the partners turn on each other.

It has long been recognized that the best marriages grow when both partners have similar backgrounds and similar reality interests. However, when the Child is in charge of planning the marriage, important dissimilarities often are ignored, and a contract which reads "till death do us part" is based on such insufficient sameness as "we both love dancing," "we both want lots of children," "we both love horses," or "we're both on acid." Perfection is seen in broad shoulders, shiny teeth, big bosoms, shiny cars, or other somewhat perishable wonders. Sometimes the bond is established on the basis of mutual protest on the mistaken assumption that one's enemy's enemy is one's friend. In much the same manner as two children, mad at their mothers, comfort each other in a bond of mutual misery, some couples hang together in us-against-the-world fashion as a protest against the malevolent "they." They hate each other's families, they hate their phony erstwhile friends, they hate the Establishment, or they hate those fatuous institutions of American "superficiality," bowling, baseball, bathing, and work. They exist in a *folie à deux* in which they share the same delusions. Yet they soon become objects of their own bitterness, and what used to be the game of "It's All Them" becomes the game of "It's All You."

One of the most helpful ways to examine similarities and dissimilarities is the use of Transactional Analysis in premarital counseling to construct a personality diagram of the couple contemplating marriage. The aim is to expose not just the obvious similarities or dissimilarities but to undertake a more thorough inquiry of what is in the Parent, Adult, and Child of each partner. A couple who enters into such an inquiry might be said to already have a lot in their favor, inasmuch as they take marriage seriously enough to take a long look before they leap. However, one of the partners, having serious doubt about the soundness of the alliance, may undertake such an inquiry on his own. An example is a young lady who was in one of my treatment groups. She asked me to schedule an individual hour for her for the purpose of discussing her dilemma over the fact that a young man she had been dating a short time had proposed to her. Her Child was immensely attracted to him, and yet there was other data coming into her computer which caused her to question whether or not marriage was a good idea. She had learned to use P-A-C accurately and asked that I help her examine this relationship on the basis of examining the P-A-C in each of them.

First we compared the Parent of each. We found she had a strong Parent, which contained countless rules of conduct and many "shoulds" and "oughts." These included the admonition that you don't rush into marriage without thinking. There were certain elements of self-righteousness, like "our kind" are the best people. It contained ideas such as "you are judged by the company you keep" and "don't do anything that is beneath you." It contained the early imprints of a home life that was highly organized, where mother was the head of the house, and where father worked hard and late at the office. There was a great store of "how to" material: How to celebrate a birthday, how to dress the Christmas tree, how to bring up children, and how to handle oneself in social situations. Her Parent came on as an important influence in her life in that the impressions had been more or less consistent. Although its rigidity was sometimes oppressive and produced considerable NOT OK feelings in her Child, her Parent nonetheless continued to be a constant source of data in all her transactions in the present.

We then turned to an examination of the Parent in the young man. His parents had been divorced when he was seven years old, and he had been raised by his mother, who indulged him in material possessions and gave him sporadic attention. She herself was Child-dominated and emotional and acted out her feelings in impetuous displays of spending, with intermittent spells of sulking, withdrawal, and vindictiveness. Father did not come through on the tape at all except as the imprint that he was a "rotten bastard, like all men." The boy's Parent was so disintegrated and fragmented and inconsistent that it did not come through in present transactions as a controlling or modifying influence over his impulsive, Child-dominated behavior. Her Parent and his Parent not only had nothing in common; her Parent also highly disapproved of his. It was readily seen that little basis existed for a Parent-Parent transaction about any subject, thus ruling out anything complementary at this level.

We then undertook an inquiry into the strength of the Adult in each and an assessment of their reality interests. She was an intelligent, well-educated young woman who enjoyed a wide variety of interests. She liked classical music, along with what was the current rage; she was well read in the literary classics; she enjoyed making things with her hands and liked to do creative, decorative things around the house. She enjoyed discussing philosophical and religious ideas and, although she could not accept the religious beliefs of her parents, did feel that some land of "belief" was important. She was reflective, conversationally adept, and inquisitive. She was concerned about the consequences of what she did and felt she had a responsibility for herself. There were certain areas of prejudice which were found to be Parent-contamination of the Adult, as "Any man over thirty who isn't married is up to no good"; "A woman who will smoke will do anything"; "Anyone who can't get through college nowadays is lazy"; "What can you expect of a divorced man?"

In contrast, her boyfriend's Adult was Child-contaminated. He continued to be self-indulgent, as he had been indulged when he was a little boy. He had been a uninterested student in high school and had dropped out of junior college in the first semester because it "didn't turn him on." He was not unintelligent but he had little interest in the serious subjects that were important to the girl. He thought all religion was phony in the same dismissing way that he thought all grownups were phony. He couldn't spell, which particularly annoyed her, and the only things he read were the pictures in *Life* magazine, the "kind of guy," she said, "who thinks Bach is a beer." He had superficial ideas about politics and felt government was bad because "it takes your freedom away." He was witty and clever but deficient in content. His primary reality interest was sports cars, about which he professed and exhibited extensive knowledge. It was apparent there was little to promote a sustained Adult-Adult relationship between the two. This level of transaction produced frustration in her and boredom in him.

We then turned to an examination of the Child in each. Her Child was hungry for affection, anxious to please, frequently depressed, and sensitive to incoming signals of criticism, which reproduced a strong feeling of NOT OK. She could not get over the fact that "someone so handsome" could fall for her. She had not had many boyfriends and had thought of herself as plain, feeling her features were so ordinary that no one could possibly recognize her after one meeting. She was swept off her feet by this fun-loving, blond Adonis, and she could not discount what a wonderful feeling it was to be loved and pursued. When she was with him she felt OK in a way she had never felt before and could not easily give this up.

His Child, on the other hand, was aggressive, self-serving, and manipulative. He had "always gotten his way," and he planned to get his way with her, too, which was part of the problem, since her Parent would not allow her to enjoy the exotic pleasures to which he proposed to introduce her. His Child so contaminated his Adult, and his Parent was so weak, that not only could he not weigh consequences, he thought the whole idea of consequences was silly and puritanical and preferred, like Scarlett O'Hara, to "think about that tomorrow."

As their relationship progressed there became less and less to talk about. Nothing existed Parent-Parent, little existed Adult-Adult, and what did exist on the Child-Child level soon produced major disturbances in the girl's Parent. The relationship then began to settle in a Parent-Child pattern with her assuming the role of the responsible and critical partner and he assuming the role of the manipulative, testing Child, reproducing his original situation in childhood.

This P-A-C appraisal was quite different from a judgment as to how "good" or "bad" each partner was. It was a search for objective data about each, with the hope of predicting what kind of relationship might be possible in the future. After much reflection on this material, the girl decided to give up the relationship as a bad deal that held little promise for happiness for either one. She was helped to see, also, how her NOT OK Child was vulnerable to the advances of men who were "less than she was" in that she had the feeling she wasn't good enough for a "really nice guy." She not only found why this relationship was not complementary, but she discovered what she truly was looking for in a man and proceeded not on the basis of her NOT OK position but on the basis of a new self-respect.

Not all relationships contrast so clearly as this one. She had a strong Parent and he had a weak Parent. There are many cases where both partners have a strong Parent, but with different and frequently discordant content. Different religious and cultural content can produce serious difficulties if each partner feels the strong need to abide by the unexamined dictates of his Parent. Sometimes this difference is glossed over in the early stage of a marriage, only to emerge with fierce urgency with the arrival of children. Although a Jewish man may agree in advance that his children be raised in the Catholic faith according to the wishes of the Catholic bride-to-be, this does not mean that he may not be deeply troubled about it later on. The feeling here is that "my religion is better than yours" and, in fact, "our people are better than your people," which soon is reduced to "I am better than you." This is not to say that differences of this kind cannot be resolved, but they require an emancipated Adult in each partner proceeding on an I'M OK—YOU'RE OK basis.

These differences are acknowledged ideally *before* the marriage. But this seldom happens. The young couple is in love; the partners, if they do have any premarital counseling, spend a perfunctory hour with the minister and then proceed on the basis of fulfilling a wish for what is called a happy marriage, frequently without the benefit of having seen one.

What are the possibilities, then, for reconstructing or salvaging a marriage which has been entered into without the benefit of this kind of analysis? Since no two people are exactly alike, the idea of perfect compatibility is illusory. The problem can perhaps best be stated in terms of comparative difficulties: It is difficult to work out the differences and make compromises, but it also is difficult to proceed with the alternative, the dissolution of the marriage. One cannot proceed on the basis of rigid absolutes, such as "divorce is always wrong," because there are other principles involved which also apply. Insisting that a woman continue to live with a cruel and abusive husband and never find happiness with anyone else is to discount the importance of human dignity in favor of retribution: You made your bed, now lie in it. To insist that a man continue to support a lazy, vengeful wife who denies any complicity in the deterioration of their marriage discounts the same principles of human dignity. This is not to say that we cannot hold to the ideal of marriage as a permanent bond, but we must not see it as a license to trap people into an arrangement in which they forever are bound by legal but no moral obligations. Sometimes people do not begin to examine their marriages until they see the divorce advancing upon them. Then the comparative difficulties begin to emerge, and they begin to comprehend the nature of the choices they must make.

A miserable marriage may make the life of the gay divorcée or the carefree bachelor seem grand indeed; yet, an impulsive choice on the basis of an unexamined assumption may lead to even further despair. That the life of the formerly married is not all it's cracked up to be is the subject of a book by Morton M. Hunt.* This author writes of the many realities following a divorce, which must be considered by people contemplating divorce in order to make their decisions on the basis of a comparison of difficulties: the difficulty of loneliness as a recurring pain, the loss of old friends who do not want to "take sides," the loss of children, the heartbreak of children, the financial ravages, the implications of failure, and the fa-

---

*M. Hunt, *The World of the Formerly Married* (New York: McGraw-Hill, 1966).

tigue of knowing one has to start all over. An Adult appraisal of one's situation must take into account these realities.

Then the inquiry must be turned to the marriage itself. Very often only one partner is willing to initiate the examination, since one of the most common marital games is "It's All You." If one partner, say the wife, comes into treatment and learns P-A-C, we then concentrate on ways in which we can "hook her husband's Adult" and interest him in learning the language as well, for only on the basis of a common language can anything begin to develop on an Adult-Adult basis. If one partner refuses to cooperate in this, the chances for saving the marriage are greatly diminished. But if both are interested enough to work at the marriage, P-A-C provides them with a tool to separate themselves from archaic Parent dictates and by now well-established game patterns.

One of the first things they may examine after they have learned the language is the marriage contract itself. The average marriage contract is a bad one, a fifty-fifty deal with emphasis on the bookkeeping. Erich Fromm calls this kind of marriage contract a "trading of personality packages." Don't they make a good match? She will be such an asset to him. Don't they complement each other? He trades his position with the Junior Chamber of Commerce for her contribution as an I. Magnined and Helena Rubinsteined "arm piece." As such, they become things and not people, in a competitive market. They must keep the fifty-fifty thing going, or the economy goes bust. This kind of contract is made by the Child. The Child has a comprehension of fairness, of fifty-fifty, but in its NOT OK position it does not comprehend a more profound principle, that of unlimited liability for another person, where one does not hold back with fifty percent but is willing to be blind to the score and give totally all the time to the partner in a community of purpose established by the Adult. In a book of exquisite meditative thought, Paul Scherer, Brown Professor Emeritus of Homiletics at Union Theological Seminary, expresses this idea in this way: "Love is a spendthrift, leaves its arithmetic at

home, is always 'in the red'. . . ."* The Child, which is a get-love
creature, cannot see love in this way. The Adult can. There is an
arithmetic of desperation in the world today, where everyone is asking for love but very few seem able to provide it. This is because of
the continuing overriding influence of the I'M NOT OK—YOU'RE OK
position of the little child. It has existed in everyone. We must keep
in mind how the little person tries to relieve himself of this burden
early in life by the original games of "Mine Is Better" and "I've Got
More." It is true that a fifty-fifty idea begins to emerge. Yet the NOT
OK seems to crowd out the idea of fairness early in life.

One morning my daughter Heidi, then four, was about to have
a treat with her playmate, Stacey. They were both preoccupied
with who was to get the bigger piece, even though they had been
reminded many times that this kind of contest only led to problems. Mother then gave them each an Oreo. It was obvious, even to
the girls, that the Oreos were identical cookies. Yet in the face of
this sameness, Heidi still could not resist the protest she had begun
and persisted, "Ha ha—I get the same as you, and you don't!" This
is the sort of hidden oneupmanship held in reserve by the Child in
the fifty-fifty marriage.

The couple seeking to save their marriage must, therefore,
enter into a collaborative effort to emancipate the Adult so that the
NOT OK in the Child, as well as the troublemaking content in the
Parent of each, can be examined to see how this archaic data continues to dominate and wreck their relationship in the present.

Common relationship wreckers are absolute declarations of
"That's the way I am—don't try to change me." Holding to a rigid
"I am a grouch before my first cup of coffee" blames a person's
faults on his nature and not his nature on his faults. The "Grouch
Before Coffee" racket ruins every morning in many families. What
could be the best part of the day, a send-off with enthusiasm to the
tasks ahead, is instead a miserable, hostile bedlam. The kids go off

*P. Scherer, *Love Is a Spendthrift* (New York: Harper & Brothers, 1961).

to school grumping, husband rushes off to work fumbling for his Tums, and mother feels let down because she has just lost her captive audience. The fact is that no one has to be a grouch before his first cup of coffee or any other time. He has a choice, once his Adult is emancipated.

An old French song goes, ". . . *l'amour est l'enfant de la liberté* ("love is the child of freedom"). Love in a marriage requires the freedom of the Adult to examine the Parent, to accept or reject it on the basis of present-day contexts and also to examine the position of the Child and the troublemaking compensations, or games, it has devised to deny, or rise above, or to throw off the burden of the NOT OK.

Married couples who enter group treatment do so for a variety of reasons. Some have heard about Transactional Analysis and come "to learn something new." Others come in search of an answer to a vague but unsettling question along the lines of "Isn't there more in life than this?" Some come because their children are having trouble. Many come because their marital relationship is in critical condition. Many of the thirty-seven couples whom I treated during one four-year period were contemplating, or at least had discussed, divorce as the only way out of their problems. Some had begun legal proceedings and were referred by their lawyers or by the Judge of the Domestic Relations Court. The crisis of seventeen of these couples (46 percent) was brought to a head with the admission of one partner to the hospital for severe depression with suicidal preoccupation or attempted suicide. Fourteen of the hospitalized patients were wives and two were husbands, and, in one case, both husband and wife were hospitalized together at their own request "to keep things even." None of these couples had been married less than ten years. All had children and some had grandchildren.

They learned P-A-C, either in a hospital group or in individual sessions in my office. When the concept was understood by both partners, there was eagerness expressed to join one of the existing married-couples groups of five couples each. The average number of treatment sessions for each couple was seventeen, roughly one session a week for four months. My married couples groups are scheduled for the last hour of the day for one hour but occasionally run longer.

Of these thirty-seven couples, thirty-five, to my knowledge, are still married, two are divorced. Four of these thirty-five dropped out of the group because they would have had to give up their games and were unwilling to do so. The other thirty-one couples report good transactions in their marriages, in which each partner is now finding the excitement of new goals, the relative absence of old destructive games, and the achievement of intimacy. In achieving one of the original goals of treatment, that of saving the marriage, we can report an 84 percent success with this group of patients.

The relationship of many couples is a complicated mesh of games, wherein accumulated resentment and bitterness have produced intricate, repeated versions of "Uproar," "It's All You," "Blemish," "So's Your Old Man," and "If It Weren't for You I Could." The rules and stereotyped plays in these games are catalogued in great detail in Berne's *Games People Play,* which has been one of the standard manuals assigned for reading to couples in treatment. These games all grow from the early childhood game of "Mine Is Better," designed to overcome the original fear of being cheated. One of the most brilliant exposés of a game existence is written by Edward Albee in the already-mentioned *Who's Afraid of Virginia Woolf?* This play illustrates that despite all the desperation produced, there still are enough secondary benefits that the games, in a sense, hold the marriage together. Some marriages are held together by virtue of one "sick" partner. If that partner begins to get well and begins to refuse to get involved in the old games, the marriage begins to fall apart. One husband, whose wife had just been released after a ten-day stay in the hospital, called me in a state of consternation saying, "My wife seems happier and better but now I can't get along with her at all." Marriage is like posture; if the shoulders begin to droop, a complementary droop must develop somewhere else to keep the head perpendicular to the feet. Similarly, if one partner changes, other changes must complement this in order to keep the relationship intact. This is one of the major weaknesses of the old types of psychotherapy, where the psychiatrist treated only one partner, often refusing to even talk with the other partner. The emphasis was on the relationship established between the psychiatrist and the patient, leaving the marriage relationship outside the door. As the patient's loyalty and behavior began to change, the marriage frequently suffered because the other partner had no conceptual tools with which to understand what was going on or to understand his own accumulation of fury and despair.

Finally, if his bank account could stand it, he entered treatment with someone else only to become further estranged as he, too, shifted the object of his affection. With little basis for communication, the way was open for new and better ways to play "Mine Is Better" in the form of "My Therapist Is Better Than Your Therapist," or "I Am Overcoming Transference Faster Than You Are" or "I'll Make a Decision About Whether to Make Love to You After My Session Wednesday." Both were indulging their Child in exclusive introspection which, though it may have provided useful data as to the origin of their own feelings, did not truly come to grips with the reality of an existence of not one person but two people in a relationship called marriage.

An item in the *Sacramento Bee,* though perhaps a bit extreme, is nonetheless pertinent: "Many a psychiatrist insists there is no mental emotional health if man does not face reality. If that be the case why do they make their patients lie on couches where it is so easy to daydream? Maybe a spiked mattress would be better."

Each partner must be willing to acknowledge his complicity in the difficulties of the marriage. The "It's All You" point of view is exposed as fallacious by Emerson in his observation that "no man can approach me except through my own act." If the husband has been abusive for ten years and the wife has taken it for ten years, then she, in her way, has participated in the exchange. If either partner refuses to acknowledge this complicity, there is little hope for change.

Arthur Miller in his sensitive story about Maggie in *After the Fall* (a character who bore a striking resemblance to his wife, Marilyn Monroe) wrote that his play was "about the human animal's unwillingness or inability to discover in himself the seeds of his own destruction."

> It is always and forever the same struggle: to perceive somehow our own complicity with evil is a horror not to be borne. [It is] much more reassuring to see the world in terms of totally innocent victims and totally evil instigators of the monstrous violence we see all about us. At all costs, never disturb our innocence. But what is the most innocent place in any country? Is it not the insane asylum? There people drift through life truly innocent, unable to see into themselves at all. The perfection of innocence, indeed, is madness.*

This "horror not to be borne" is understandable when one considers that the admission of complicity adds still another load to the crushing NOT OK burden which has caused the problem in the first place. The admission of guilt is hard. It is this final affront to the abject Child, this additional burden, that is referred to by the German theologian Dietrich Bonhoeffer: "Is not this to lay another and still heavier burden on men's shoulders? Is this all we can do when the souls and bodies of men are groaning beneath the weight of so many man-made dogmas?"†

Understanding **Structural Analysis**—the nature of the Parent, Adult, and Child—shows us a way out of this dilemma of on the

---

*A. Miller, "With Respect for Her Agony—but with Love" *Life* 55:66 (Feb. 7, 1964).
†D. Bonhoeffer, *The Cost of Discipleship* (New York: Macmillan, 1963).

one hand the impossibility of change without the admission of complicity, and on the other the crushing implications of the admission of guilt. In a very practical way we can see a difference in how we confront a person with what he does. If one says, "You are a cranky, ill-tempered, difficult, unpleasant person, and this is what is wrong with your marriage," one is simply supporting the NOT OK position and producing feelings that make the person even more cranky, ill-tempered, difficult, and unpleasant. Either that, or you drive him into a deepening depression. If, on the other hand, one can sympathetically say, "It is your NOT OK Child that constantly makes trouble for you and comes on in the old cranky and ill-tempered way to destroy your chances for happiness in the present," there is some objectification of the dilemma, and the person sees himself not as a total zero but as a combination of past experience, both plus and minus, which produces difficulty. Moreover, it makes possible a choice. A person can acknowledge this reality about himself without falling apart, and this acknowledgment can begin to strengthen his Adult for its function of examining the Parent and the Child and the way in which these old tapes come on to produce the tyranny of the past.

Without the acknowledgment of "my part in our problems" Transactional Analysis or game analysis can simply become another way of expressing hatred: "You and your damn Parent," "Your nasty Child is coming on again, dear." "There you go, playing a game again." These constructions then become clever and abusive epithets in a new game of "game calling." As we see the problems that can arise, we begin to understand the significance of the idea expressed in the title of Arthur Miller's article about his play, "With Respect for Her Agony—but with Love."

A commitment to this idea is what is required of couples entering treatment if they are to succeed in building something of value in their marriage. A final question emerges: When we stop playing games then what do we do? What else is there? What do you do with an emancipated Adult?

The Establishment of Goals

A ship with no destination drifts and is carried along by the prevailing tides, now up, now down, groaning and creaking in the high seas, tranquil and lovely in the calm. It does exactly as the sea does. Many marriages are like this. They stay afloat but they have no direction. The priority input in their decision making is, What are other people doing? They conform to their social circle in attire, housing, raising children, values, and thinking. "As long as others are doing it, it must be OK," is their standard of what to do. If "everyone" is buying a certain kind of luxury automobile, they also will buy one, even if their time-payment coupon books already constitute a library of monthly bad news. They have not built their own set of independent values concerned with their own particular realities and therefore frequently end up disillusioned and in debt.

Only the Adult can say "no" to the Child's clamoring for something *bigger, better,* and *more* in order to feel more OK. Only the Adult can ask the question, If four pairs of shoes make you happy, will ten pairs make you happier? The rule is that each increment of material possessions brings less joy than the one that immediately preceded it. If one could quantify joy, it is likely that a new pair of shoes brings more happiness to a child than a new car brings to a grown man. Also, the first car brings more joy than the second, and the second more than the third. H. L. Mencken said, "A man always remembers his first love. After that he begins to bunch them."* The Child in us needs bunches—as on Christmas morning: surrounded with gifts the child cries, "Is that all?" A little boy was asked on a children's television program what he got for Christmas. "I don't know," he said, distressed, "there was too many."

An Adult examination of a family's realities can weigh whether

---

*H. L. Mencken, *The Vintage Mencken,* gathered by Alistair Cooke (New York: Vintage Books, 1956).

or not the acquisition of a certain possession will be worth (in terms of joy) the mortgage, the department store bill, or the diversion of the money from something else. The Adult can also give in to the Child's need to collect bunches of possessions by taking up a hobby such as collecting stamps, coins, rare books, model railroad equipment, bottles, or rocks. The Adult can determine whether the expenditure for these collections is realistic. When it is, the "bunching" is fun and harmless. If it is bankrupting the family, however (e.g., collecting villas, sports cars, and original Picassos), the Adult may have to say "no" to the Child's fun.

Decisions regarding hobbies, possessions, where to live, and what to buy must be made according to a set of values and realistic considerations unique to the marriage. Agreement about these decisions is extremely difficult if goals for the marriage have not been established. A couple in treatment may learn to see the difference between Parent, Adult, and Child, but they are still on the same social sea, and if they do not chart a course, they will, despite all their insights, continue to follow the old ups and downs and fun and games. It takes more than knowing something to muster the power to cut through the social currents. It takes the establishment of and embarking upon a new course in the direction of goals arrived at by the Adult. Persons either set a new course or they fall back into the same patterns of drift. It does not matter how many charts they have.

This is where the considerations of moral values, of ethics and religion, become important to the course of a marriage. A man and wife must undertake some fundamental inquiries about what they consider important in order to chart their course. Will Durant views the fundamental problem of ethics in the form of the question, "Is it better to be good or to be strong?"* This question can be asked in many ways in the context of the marriage. Is it better to be kind or to be rich? Is it better to spend time with the family or

---

*W. Durant, *The Story of Philosophy* (New York: Simon & Schuster, 1926).

to spend time in civic activities? Is it better to encourage your children to "take it on the chin" or to "hit back"? Is it better to live big today or hoard every penny in the bank for tomorrow? Is it better to be known as a thoughtful neighbor or to be known as a civic leader?

These are questions which can lead to hopeless forensic entanglements unless they are asked by the Adult, for they are difficult even then. It is not enough to know what opinions the Parent in each partner contains in answer to these questions. It is not enough to know the Child needs and feelings of each. If the Parent or Child data is in disagreement, there must be some ethical standard accepted by both, which can give direction to the course of the marriage and value to all decisions that must be made. It has been said that "love is not a gazing at each other, but a looking outward together, in the same direction." The Parent and Child in each partner may lead backward to great divergence. Only through the Adult is convergence possible. Yet the goal "out there" cannot be established without moral and ethical considerations. One of my frequent questions to a couple in an impasse over "what to do now" is: "What is the loving thing to do?"

This is the reaching beyond scientific evaluation to the possibility of the commitment to something better than what has been before. What is "being loving"? What is love? What kinds of words are "should" and "ought"? These questions are considered in depth in Chapter 12, "P-A-C and Moral Values."

P-A-C and Children

Those who cannot remember the past
are condemned to repeat it.

—George Santayana

The best way to help children is to help parents. If parents do not like what their children do, it is not the children alone who must change. If Johnnie is a hot potato, he is not going to cool off by being thrown from expert to expert, unless something is done about the oven at home. This chapter is written to help parents help children. "Experts" cannot do the job parents can.

It is true there are many professional child-raising experts, including child psychiatrists and child psychologists who perform testing and do treatment. In England the baptism of a child is referred to as "having the child done." In similar vein it may seem that bringing Junior to a child psychiatrist may imply having the child "redone" or perhaps "undone." Unless the parents are simultaneously being redone, I believe most of these efforts are a waste of time and money. I believe most parents intuitively feel the same way, but some parents, not knowing what else to do, or not wanting to become involved themselves, go along with the idea of child treatment, if they can afford it. Many other parents shy away from the unknowns of getting help with child raising, viewing their situation as some kind of Pandora's Box that perhaps is best left unopened. They read the latest books, consult the newspaper columns, and play "Ain't It Awful" over morning coffee. They practice "patience" without answers in the hope that Junior is "going through a phase" and base their hopes on an uncertain principle that permissiveness is a good thing. The answers they seek are not forthcoming, and they struggle through child raising with the small comfort, "Well, at least I'm bigger than he is." Some parents exercise their "bigness" violently, battering and bullying their children into shape. Then the day of reckoning arrives, somewhere around adolescence, when "he's bigger than I am." There is great misery all around, for the parents and for the children. This does not have to be. It is the purpose of this chapter to brighten the picture of child raising with the application of P-A-C, not only to the relationship between the parents and children, but also to that between children and other children.

The psychiatric treatment of children is a relatively recent development. While the early psychoanalytic theorists emphasized the significance of what happened to the child in the early family setting, working directly with children was not a part of the early application of that theory to treatment. One difficulty was the problem of communicating with the little person. The other was the early recognition that little could be accomplished in working with the child without the involvement of the significant grownups in his environment, mainly his parents.

The first comprehensive clinical structure for treating children developed in the 1920s in what became known as the Child Guidance Clinic. This developed into a conjoint "treatment experience" for parent and child, with the child being "treated" by a method called play therapy and the parent being helped with social case-work counseling. Central to the method was the opportunity for parent and child to "express feelings," with the aim of eliminating a potent source of provocation to negative and destructive behavior. Through the use of toys and other symbolic media of communication the child was encouraged to turn on his tormentors, the parents, with a cleansing catharsis of "negative feelings." Thus when Junior flushed mother doll down the toilet or broke the arms off the little sister doll, notes were made for the next session of "conferencing," an activity of great importance to the clinic staff. The assumption was that these expressions would clear the way for the development of more positive feelings based on the new insights the parents would derive from their work with the social worker—that after a given number of "I hate you's" the "I love you's" would somehow follow. Yet, inadequate understanding by the parents of the actions, or transactions, which produced feelings frequently left the situation unchanged. In fact, the situation often got worse as the child was told "expressing feelings is a good thing," turning the family into a battlefield, with Junior the commanding general. It was like nose drops. It relieved the congestion for a while but was not particularly helpful in preventing tomorrow's congestion. Some people go through life expressing feelings. In both instances, it would appear that the activity is at the wrong end of things. It is not that expressing feelings or using nose drops does not have certain therapeutic benefits, but there is more to it than that.

The emphasis in these early modes of treatment was on what the *child* could achieve and how his behavior could change, although there was some acknowledgment that the parents had to be involved. Our emphasis in Transactional Analysis is on what the *parents* can achieve so that the nature of the transactions between parent and child will change. When this happens, the change in the child will soon follow.

Everyone recognizes the increasing complexities of the culture and social structure in which we live today, with the many pressures that tend to weaken and even destroy the family as the primary social structure for meeting the emotional needs of children.

Under the impact of uncertainties, the outpourings of news and entertainment media, and a flood of demands, the modern mother feels embattled and frequently on the verge of disintegration in her struggle with frustration. Everything around her is in conflict. Her sensitivity is dulled, as it has to be, as within seconds her television set moves from the ghastly reports of war to the glories of a new life with Clairol. Her Parent is in conflict with her husband's Parent over the finer points of the Little League. Her Parent rides her Child in an internal dialogue that makes her feel a failure as a mother. Her children scream at each other and at her. She reads to get more data, but the data conflicts. One authority says "spank," another says "Never spank," and still another says "spank sometimes." In the meantime, her feelings build up to the point where she wants to "beat hell out of the little bastards." Her house is full of appliances which help her do everything with the greatest of ease. But what she needs the most is a tool to bring order out of the chaos, determine which goals are important and which are not, to find realistic answers to the repeated question: How can I raise my children properly?

To this question Grandmother might observe sagely, "We didn't have all this trouble back in the good old days before all these books on modern psychology." Grandmother has a point there, since there was a great deal of good in the old days. Gesell and Ilg observed:

> In the more olden times, the world of nature and of human relationships expanded in a rather orderly manner, keeping pace with the maturity of the child. The home was large, the membership of the family numerous, and usually there was yet another child to be born. Someone was always near to look after the preschool child and to take him by graduated stages into his widening world, step by step, as his demands gradually increased. There was free space around his home, a field, a meadow, an orchard. There were animals in barn, pen, coop, and pasture. Some of these fellow creatures were young like

himself. He could feast his eyes on them, touch them, sometimes even embrace them.

Time has played a transforming trick with this environment.
The apartment child, and to some extent even the suburban
child of today, has been greatly deprived of his former compan-
ions, human and infrahuman. Domestic living space has con-
tracted to the dimensions of a few rooms, a porch, a yard;
perhaps to a single room, with one or two windows.*

They lament the loss to the small child of today of "ample intimate
contact with growing life, with other children, with a variety of
adults."

Not only does there seem to be a lack of these early good experiences, there is also a deluge of frightening incoming data. It is
true there have always been wars and atrocities, but they didn't
happen in the living room on the television set. Long before the
child is able to cope with the elementary difficulties of getting along
in the family, he is introduced to what my little girl calls "a screwy
world" of race riots, child prisoners blindfolded at bayonet point,
mass murder, and world leaders debating the possibility of global
annihilation. Add to this the difficulty for the small child of sorting
out what is fact and what is fiction: Is this the news or is it a movie?
Is that the head of the cavalry or is that the governor? Does smoking cause cancer or is it the breath of spring?

During the Cuban crisis in 1962 my daughter Heidi, then in
kindergarten, where the youngsters were being taught "atom bomb
drills," said to her mother, "Mama, let's talk about the war and the
bomb and things." Mother replied, "OK, Heidi, what shall we say
about it?" To which Heidi replied, "You say all the words, Mama.
I don't know any of the words about this."

This, then, is the world as we find it, not some sheltered pastoral scene with little lambs and yellow flowers and the *Good Ship*

*Arnold Gesell and Frances Ilg, *Infant and Child in the Culture of Today* (New
York: Harper, 1943).

Lollipop, but a world of anger and clashing sounds, amplified to such levels that the temptation is to turn it off and not care about the difference between Clairol and crime, or the difference between the assassination of a President and the comic demise of a cattle rustler in a black hat.

Will Rogers once said, "The schools ain't what they used to be and never was." Maybe the good old days "never was" either, but the badness did not touch children as early and as intimately as it does today. This does not change the problem, but it makes it more urgent than ever that parents have a tool to help their children develop an Adult early to get along in the world as it is.

Where to Start

Ideally, we would like to start at the beginning. One effective application of Transactional Analysis has been a teaching program for expectant parents, which has been conducted in Sacramento since 1965 by Dr. and Mrs. Erwin Eichhorn. He is an obstetrician-gynecologist and she is a teacher of nursing at Sacramento City College. In most obstetrical practices preparation for childbirth generally includes instructions for the parents-to-be, particularly the mothers, in what to expect during pregnancy, labor, and childbirth, with information as well about the physical care of the infant. This frequently is supplemented with various books and films, which have portrayed an idyllic life with the the New Baby. There may be some discussion given to the negative aspects of the experience, such as the possibilities of "post-baby blues," or fatigue, or colic, but there has rarely been any specific examination in depth about the relationship between husband and wife, new mother and new daddy, and this beautiful and sometimes terrifying new little person, the baby. Most obstetricians would have liked to help the young couple in this way, but there was no system that could be quickly taught, easily understood, and readily put to use. Many an obstetrician has spent many an hour in sympathetic discussion of difficulties in the family situation, relieving anxiety by answering questions, and attempting to allay fear by kindly support. Others have relied on a more parental position, which in essence says, "You follow my instructions and do as I say and everything will turn out all right." However, if there are serious relationship problems between the couple, this kind of approach may relegate these problems somewhere down on the priority list, since, after all, the baby must come first. Yet, unresolved, they remain a source of constant irritation and alienation to both mother and father during the earliest months or years of the infant's life, during which time the most fundamental imprints are made on the child.

The Eichhorns, who are both members of the Board of Directors of the Institute for Transactional Analysis, began introducing the teaching of P-A-C to their expectant-parent classes in 1965. Evening meetings for both husband and wife are scheduled weekly. Attendance is voluntary, but most couples attend regularly. In addition to the regular instructions about pregnancy, labor, and delivery, the fundamentals of Transactional Analysis are taught. These are taught in terms of the actual experience the couple is involved in—having a baby. It is a tool that is being put together for a special purpose, but one which the couples find can be used for many other problems in living after the baby arrives. Total group instruction in P-A-C for each couple is about twenty-four hours, but the language thus developed provides the basis for further discussion as the mother-to-be arrives for her regular obstetrical checkups, frequently accompanied by her husband, who is made to feel a part of the show rather than an onlooker.

It was found that an understanding of P-A-C early in the pregnancy helped the couple understand the source of some new, rather complicated, not-all-positive feelings. Young people whose Parents contain many urgent, undifferentiated recordings about intercourse and pregnancy should not be surprised that these recordings replay in this emotionally charged experience. A young couple, even if they have planned for and eagerly await pregnancy, do find they are subject to periods of "unexplainable" depression. A marriage license and a little white cottage do not erase the old Parent tape wherein "I'm pregnant" would be horrible news indeed. Nor does it change the Parent tape for the husband which comes on in the old way to his realization that "I got you pregnant."

There are many other intense feelings associated with pregnancy, which Gerald Caplan refers to as "a period of increased susceptibility to crisis, a period when problems of an important nature appear to be present in an increased degree."* In addition to the external economic and social changes there are internal changes, both metabolic and emotional. For mother there is a new role, particularly if this is her first baby; there is the aloneness of labor and the loneliness of being home with the baby, particularly if she previously had been a career woman; and there is the new responsibility of structuring time. There also is a profound realization for the woman having her first baby that she will never be a little girl again, that she has passed beyond the pale into the older generation: She now is a mother. This is the same kind of sentiment over the shortness of life and the irrevocable passing of time which makes people cry at weddings. The sacramental moments in life, while opening doors to the future, close the door on the past, and there is no turning back. These are the feelings of the young mother, too.

Sometimes these feelings become so depressive that there develops what is called a post-partum, or after-birth, psychosis. In these cases the Child becomes so overwhelmed that the boundary breaks down, allowing for complete contamination of the Adult. Mother cannot handle her own overwhelming needs and is totally incapable of caring for her baby.

One patient, whom I saw first in an acute episode of post-

*G. Caplan, *An Approach to Community Mental Health* (New York: Grune and Stratton, 1961).

partum psychosis following the birth of her first child, was able to leave the hospital three weeks after being introduced to P-A-C. She was able to assume the care of her baby, and her Adult grew stronger as she continued coming to group treatment. The real test of this strength came at the time of her second pregnancy two years later. Knowing what she had been through before, she proceeded through her pregnancy with a great deal of apprehension. But she was able to discuss this apprehension in P-A-C terms with her obstetrician. (The fact that two doctors, one an obstetrician and the other a psychiatrist, both talked the same language was in itself reassuring.) She delivered her child and remained in good spirits throughout the post-partum period. (It is not uncommon for the post-partum psychosis to recur with every pregnancy.)

These then are some of the feelings which can be understood and overcome by P-A-C. As husband and wife are able to use their newly acquired language, they both share in the excitement ahead. Eichhorn reports that when the doctor "comes on" Adult, it is easier for the husband to become a father. The Parent-Child relationship between some obstetricians and their patients essentially excludes the father. Mother and doctor seem to be involved in an activity in which they claim exclusive expertise, and husband is left to pace in the waiting room. Most modern hospitals allow the husband to be with his wife and assist her during the hours of labor, and some, but not all, allow him in the delivery room. Eichhorn reports that in his practice the father-baby system gets going early. The husband becomes involved in what he can do to help during labor, how he can massage and relieve physical stress, how he can protect his wife from the loneliness of labor, and how, in fact, she can rely on his Adult even if, in her own fatigue and apprehension, her Child takes over. When a couple comes through a crisis like this together, the meeting of any other crisis in life has a precedent. "If we did this, we can do anything!" These fathers quickly talk about "our" baby. Both mother and father feel good about themselves, and this is transmitted to the infant.

These fathers are helped to recognize early in the pregnancy, as does Caplan, that

> the pregnant woman needs extra love just as much as she needs extra vitamins and proteins. This is especially so in the last few months of pregnancy and during the nursing period. During pregnancy she often becomes introverted and passively dependent. The more she is able to accept this state, and the more love and solicitude she gets from the people around her, the more maternal she can be toward her child. Professional workers cannot give her the love she needs, but they can mobilize the members of her family, and especially her husband, to do so. In our culture, husbands and other relatives are often afraid of "spoiling" the expectant mother and special efforts are needed to counteract this attitude.*

Being together for the delivery itself is an ideal culmination for the couple that has been prepared; but even if the couple is apart during the delivery, their introduction to P-A-C not only has helped them through the time of pregnancy but also allows for the maximum freedom from conflict that is essential to the early nurture of the new infant. The warm, stroking mother is the mother who is free from the internal Parent-Child dialogue which evokes the NOT OK in herself. Her emancipated Adult can hear the facts, can dismiss the "old wives' tales," and can react to the spontaneous maternal feelings of wanting to hold and pet and caress without checking first to see if it's all right. One of the most commonly expressed Parent ideas in the expectant parent groups is that "you shouldn't pick up a baby all the time because you will spoil him." If this tape comes on every time the new mother goes to stroke her child, it is clear there exists a conflict, which somehow will get through to the child. The Adult in the mother can examine this bit of dogma and proceed with her own estimate of the matter, which

*Caplan, *An Approach to Community Mental Health.*

would come closer to: If you baby a baby when he's a baby, you won't have to baby him the rest of his life. (These terms, "spoiling" a child or "breaking" a child of habits, have always seemed to me so crude and cruel when applied to human beings that surely they were invented by some wicked fairy-tale step-mother who lived in a dark, dank tower somewhere on the moors!)

The mother with a strong Adult can handle the frequently loaded grandmother or mother-in-law situation in a way which will minimize the devastating crossed transactions. She can appreciate that grandmother has a P-A-C, too, and both P and C are vulnerable to being hooked. Or her Adult may tell her mother-in-law that they are having a maid in to care for the house and that she, the mother, will take care of the baby. Her Adult can let the dust collect while she tends to the infant, even if her rich Aunt Agatha is arriving with a present that very night. In short, the new mother and father have a choice in how they will proceed in bringing into existence this new precious unit, their family, which has a new baby and a new father and a new mother.

One of the most helpful understandings in the raising of small children is an awareness of the I'M NOT OK—YOU'RE OK position. The child remains afloat by virtue of the mother's OK. He feels NOT OK, but as long as she is OK there is something he can hold onto. The value of a parental stroke for the child is exactly proportionate to the value the child sees in his parent. It is readily apparent that when Mother's Child gets hooked and she gets involved in a Child-Child slugfest with Junior, he senses his world is in bad shape indeed. On the one hand is a NOT OK Child and on the other hand is a NOT OK Child. If this kind of transaction predominates in the early life of the little person, the way is open for the establishment of the position, I'M NOT OK—YOU'RE NOT OK, or, in the extreme, I'M OK—YOU'RE NOT OK.

The mother and father (particularly the mother, since she is the most influential parent in the early years) must be sensitive to the NOT OK Child in themselves. Until such time as parents, partic-

ularly mothers, develop the necessary sensitivity, perceptual powers, and interest to apply such a tool as P-A-C to child rearing, we
can expect that the malignancy of the NOT OK position will spread
and grow worse. If the Child in the mother has a strong NOT OK position, and it is easily hooked by such life hitches, or obstacles, or
disappointments as the obstinate behavior of a small child who also
has a NOT OK Child, the way is open for a take-over by the Child in
the parent, which triggers a regressive sequence of events with more
and more archaic circuits taking over in a screaming game of "Mine
Is Better" with Mother winning the final round with "I Am Bigger."

It is easy to see that only through the Adult can the little person learn more effective ways of living. But children might well ask:
How do you develop an Adult without having seen one? Children
learn by imitation. One of the most effective ways a small child can
develop his Adult with increasingly strong circuits of control is to
have the opportunity to observe his parent, when the parent's
Child obviously has been hooked and is struggling to take over
with an angry outburst, control his Child and keep his response
Adult, which is to say, reasonable and considerate.

Demonstrating what an Adult is is far more effective than
defining what an Adult is! This raises the question as to whether or
not parents should teach their children P-A-C. Judging from reports of P-A-C-schooled parents of young children, the child can
understand the rudiments of P-A-C at a surprisingly early age,
three or four. This can come about through the child's exposure to
the Transactional Analysis of his parents. When parents are involved in analyzing a transaction and are doing it with obvious rel-
ish, the small child gets the meaning of what is taking place. Many
parents of three- or four-year-olds have been startled to hear the little one come through with a remark in which he used the word
Parent and Child correctly.

When a five-year-old says, "Dad, don't use up all your Parent,"
he conveys the understanding that Dad has "parts" too, that he has
a Parent and Child that can be hooked. When Dad says to the five-

year-old, "If you keep doing that, you're going to hook my Parent and then we'll both feel bad," the way is open to an Adult-Adult acceptance that both the little person and his father have feelings and can be pushed too far. This Adult-Adult position cannot possibly develop if Father roars, "You do that again and I'll slap you silly!" All this serves to do is to shut down the computer in the child; he cannot ponder the pros and cons of "what he was doing," only the fact that he will be slapped silly. And so endeth the lesson. Father probably heard it that way from his father, ad infinitum.

A word of caution is important here! Any reference to P-A-C (particularly game calling) by parents, when their youngster is coming on Child, is heard as Parent. In short order, the whole idea can become Parent, impairing its usefulness as a tool in producing Adult-Adult transactions in the home. You can't teach P-A-C to an angry, adrenaline-charged child. The answer is to *be* Adult when the heat is on. P-A-C can be talked about academically on other occasions, giving the youngster the data by which he may be able to come to his own "a-ha" experience: Hey, that's what *I* do! In time the use of these words enables children to begin to express feelings with words rather than to act out their frustration in temper tantrums in order to dominate the situation with the only tool they have had, their emotions.

When one considers the almost insurmountable barriers to the development of the Adult in childhood, the amount of irrationality or just plain cussedness prevalent today is not surprising. The child's curiosity, his need to know, is a manifestation of his developing Adult and should have the protection and support of sensitive, perceptive parents. However, sensitivity and perceptiveness can be hard to come by for couples who find the insistent demands of their children too much to cope with because of the prior demands of their own Parent and Child. The emancipation of the Adult from archaic data can make such positive attitudes as patience, kindness, respect, and considerateness a matter of choice. The choice is one of being consistently helpful to the small child or beating him down and back into catastrophic terror with the bellowing of the archaic Parent, a product of countless generations of parental self-righteousness.

As the philosopher is conditioned to ask in every transaction, "What follows?" so the parent may find useful the reflex to ask, "What came before?" What was the original transaction? Who said what? Children's responses are not far removed from what stimulates them. A practice in asking the right questions, and listening to the answers, will get to the source of difficulty quickly. If a youngster comes to Mother in tears, she has two jobs. One is to comfort the upset Child and the other is to get the Adult working. She can say, "I can see that someone made you feel badly . . . and it's hard to be little . . . and sometimes the only thing you can do is cry . . . can you tell me what happened? Did someone say something that made you feel bad?" Very quickly the transaction that produced the trouble is reported, and the mother and youngster can talk about it Adult-Adult. Sometimes we find children taking advantage of each other. For instance, big sister trades her nickels for little sister's dimes because "nickels are bigger." We promptly chastise big sister for this highhanded deal, but we must ask ourselves, "Where did she learn that?" It may have been just innate inventiveness, or it may have been a lesson learned from Mother and Dad: Be clever and make lots of money; it's really more important than people (even little sisters).

We often forget how quickly the value judgments we make are reflected in the actions of our children. H. Allen Smith recounts a story written by a nine-year-old girl: "Once upon a time there was a little girl named Clarissa Nancy Imogene LaRose. She had no hair and rather large feet. But she was extremely rich and the rest was easy."

Besides asking "What came before," the Adult can also ask what is *the* important consideration here? The Parent is effusive and can replay all kinds of reasons why one must, should, mustn't etc. This diatribe hits the little child as if it were coming out of a fire hose, and he hears nothing at all. The Adult can be selective and present the best point, not all points.

A transaction particularly disorganizing to a youngster is one in which the parent, in answer to a request by the youngster, tediously gives *all* the reasons why he shouldn't do something instead of simply stating *the main reason.* If that main reason isn't robust enough to be stated in simple terms, then perhaps it should be rejected.

The six-year-old comes into the kitchen followed by four playmates. The time is 4:45 in the afternoon. Mother is preparing dinner, which she also is sampling. The six-year-old says: "Mother, may we have something to eat?"

Mother responds, through a mouthful, "No, it's almost suppertime. You eat too much candy. It's bad for your teeth. You'll have to get fillings. [She has fillings.] If you eat now, you won't eat your supper. [She is eating now.] Got on out and play. You're always dirtying up the kitchen. Why don't you ever put anything away?" This is a splendid opportunity for Mother's Parent to torment the youngster with a whole series of moralizing "moreovers." The children grump and leave, and are back in ten minutes for more "fun and games."

The *real* reason Mother was annoyed was "why must you always bring all the neighbor kids in? I'm tired of giving all the popsicles to the neighbor kids. There are never any left for us." At this particular moment this was the real reason, and it was a valid reason. But being unable to come on straight, she loaded her daughter down with an avalanche of peripheral data. Rather than growing from this type of a transaction, the child shrinks and begins to learn peripheral (or distorted) ways to beat the Establishment. If "politeness" prevented her mother from giving the real reason, she would have been better off simply to say, "No—we'll talk about it later." Then, in the absence of the other children, she could have explained the realities to her own child. Or she could have devised a treat that would include the other children but wouldn't use up the "expensive" treats, the popsicles.

As it was, she loaded the transaction with inconsistencies, producing questions in the child's mind: How come you're eating and we can't eat? What's wrong with fillings? You have fillings. You dirty up the kitchen, too. You eat candy. How much is "too much candy"? It is just as oppressive, if not infuriating, to the child as it would be if a grownup, on asking his boss for a raise, were made to listen to the full reading of the Ten Commandments.

In proving any point the successful man presents the *best* evidence. He doesn't clutter up his case with irrelevancies. The same rule applies to parents. They are successful in discipline if they stick to the *best* reason. This gives the child's Adult something solid to process, and his computer isn't loaded down with inconsequential data. He also has the opportunity to come out of the transaction with self-respect instead of an overwhelming NOT OK. You refrain from reading your employee the Ten Commandments because you respect his Adult; if you want the Adult in your youngster to grow, you must also respect him.

The School-Age Child

When the five-year-old manfully strides up the walk on that celebrated first day of kindergarten, he takes along with him about 25,000 hours of dual tape recordings. One set is his Parent. The other, his Child. He also has a magnificent computer that can click off responses and produce brilliant ideas by the thousands, *if* it is not totally involved in working out the problems of the NOT OK. The bright little boy is the one who has had lots of stroking, who has learned to use and trust his Adult, and who knows his Parent is OK and will remain that way even when he feels NOT OK. He will have learned the Adult art of compromise (although relapses may be expected), he will have the confidence that grows from success-ful mastery of problems, and he will feel good about himself. At the other extreme is the shy, withdrawn little boy whose 25,000 hours of tape recordings play in a cacophony of shrill supervision and criticism to the low, steady rhythm of NOT OK, NOT OK, NOT OK. He also has a magnificent computer, but it has not had much use. Luigi Bonpensiere, in a remarkable little book about piano playing, com-mented on how poorly we use the superb physical apparatus of the human body: "It is like having the most perfect apparatus of preci-sion, planned and built for a highly efficient operator to use, and then relinquishing it to a poorly trained engineer, who, in the end, will complain of its limitations."*

If the child cannot use his computer, most probably it is be-cause either he has never seen one used, or he has had no one to help him learn how to use his. If he does poorly in school, his com-plaint of his limitations will be expressed, "I'm stupid," and his parents' statement will be, "He's not working up to his potential." The basic problem is the severity of the I'M NOT OK—YOU'RE OK position. School, unless it has truly competent teachers, is the place where scholastically the "rich get richer and the poor get poorer." In a child who obviously has a school problem—disrup-tive behavior, daydreaming, or poor achievement—one can expect to find the I'M NOT OK—YOU'RE OK a matter of continual preoccu-pation. School is a competitive situation with too many affirming threats to the Child and too few opportunities, at the start, for even token achievements to minimize the NOT OK. The early school years can be the beginning of a pattern of recurring testing transac-tions which, as he feels it, underline the reality of his NOT OK posi-tion with associated feelings of futility and despair. The really urgent aspect of this situation is that all of life is competitive, be-ginning with life in the family and extending through all of school

---

*L. Bonpensiere, *New Pathways to Piano Technique* (New York: Philosophical Li-brary, 1953).

and into the grownup world of life in society. Throughout life the feelings of, and the related techniques for, coping with the NOT OK position, which the youngster establishes in the family setting and in school, can persist into the grownup years and deny him the achievements and satisfactions based on a true sense of freedom to direct his own destiny.

My advice to parents of a youngster who is having difficulty in school is to learn P-A-C, to take it seriously, and to begin to handle their transactions with their youngster, Adult to Adult, with therapeutic assistance if necessary. They must always keep in mind the primary influence of the NOT OK. The rule is: When in doubt, stroke. This will keep the frightened, anxious Child comforted while the Adult gets on with the realities of the situation. Very often, however, these realities are not made clear to the child. Dr. Warren Prentice, Professor of Education at Sacramento State College and a member of the Board of the Institute for Transactional Analysis, suggests that a child who brings home a paper marked "Try a little harder" interprets this as a parental, undifferentiated YOU'RE NOT OK. What he needs to know is "try *what* harder." The statement "too slow" implies the question, How fast would be about right? Prentice says the child needs to be helped to identify areas in which he is or can be successful, and this is not done by a written test, since this medium itself evokes the old tape, "I can't do it, so why try?" It is done by listening to and talking with the child. He says if a child is having trouble in school, it is pointless to assume that more of the same in summer school or on weekends is going to help him unless a specific problem is isolated and met. The Parent says, "Do more." The Adult asks, "Do *what* more?"*

This is reminiscent of an editorial that appeared in the Kansas City *Star* referring to a certain public official who had declared that there were too many minors in the beer taverns. Stated the edito-

*The Center Circle, Newsletter of the Institute for Transactional Analysis, Vol. 1, No. 7 (October 1967).

rial: "He says there are too many minors in the beer taverns, but as usual he fails to point out how many minors would be about right."

Following an address on Transactional Analysis to a group of educators I was told, "We've got to get this into the schools." I certainly agree. Many parents agree, too. The question "Should Transactional Analysis be taught in the schools?" was asked of sixty-six parents who had just completed an eight-week Transactional-Analysis lecture series. In response 94 percent of the parents said yes for high school and 85 percent said yes for junior high school and elementary school.

Education is heralded as the greatest medication for the ills of the world. Those ills, however, are deeply embedded in behavior. Therefore *education about behavior* through an easy-to-understand system like P-A-C could well be the most important thing we can do to solve the problems which beset us and threaten to destroy us. The task involved is almost beyond comprehension; yet in some way and at some point we have to make some kind of break in the relentless march of the generations toward insanity or other forms of self-destruction which originate in childhood.

Preadolescents in Treatment

Preadolescence is viewed by some parents as the time of the last-ditch stand before the hormones and hairdos of adolescence appear, complicating what may already be a difficult relationship between youngsters and their parents. This is a time when youngsters are receiving maximum exposure to new ideas of the world about them, in school and in social contacts. It is a time when youngsters implement their early games by new inventive moves, which drive some parents to distraction and others to the doctor. We must keep in mind that the Child needs the security provided by relatedness, consistency, stroking, recognition, approval, and support. Some children have found that the successful way to achieve this security is to conform and cooperate and, if their parents allow it, create. Others who have not learned to get stroking this way will continue to use the early manipulative techniques of the three-year-old, such as acting out, testing, rivalry, evasion, stealing, and seduction. These techniques can be disorganizing to the family, particularly when the preadolescent applies his sharp mind to their perfection.

In 1964 I started a group for preadolescents, nine to twelve years old. The group met once a week. A group for their parents met every two weeks in the evening. These groups continued throughout the school year. At the end of the year each child, with his parents, was invited to come in for a survey of results. The changes were striking. Even the physical appearance of most youngsters had changed; many children wear their NOT OK in their facial expression and posture, and there was noticeable improvement in both. All families reported an improvement in communication. The youngster felt he could talk about his feelings and explain his point of view without provoking a parental storm or a sulky impasse. Parents discovered they were able to make realistic demands and impose realistic limits without provoking a siege of acting-out behavior. The preadolescents and their parents were urged to use the "contract" concept, which is a statement of mutual expectations drawn, discussed, and restated from time to time at the Adult-Adult level. Where the contract was clear, where it contained the do's and don'ts as well as the consequences of the broken contract, the relationship between parent and child improved markedly. The contract is one of the best instruments I know of for assuring consistency in direction and discipline; yet, because it is drawn by the Adult, it can be re-examined from time to time by the Adult with the further benefit of keeping it up to date and flexible enough to meet changing realities. Many parents treat their preadolescents in the same way they treated them when they were four years old. Often it is because they wish to maintain strict parental control, but more often it is simply because they do not appreciate how much the child changes from year to year and increases his ability to use his Adult. It is, after all, with the Adult that the young-ster learns realistic inner control. The appreciation that he does have an Adult and is not still just a "stupid little kid" immediately takes a great deal of the friction out of family transactions.

My preadolescent patients learned P-A-C quite easily and found it exciting and useful. Backed up by the approval of interested and concerned parents, their understanding of Transactional Analy-sis developed rapidly. As the internal *and* external Parent-Child di-alogue became less critical, there was a freeing up of the Adult to go to work on the important business of finding out about life. This is the time when boys and girls daydream about what they want to be, when they begin to develop intense idealism and feel a new close-ness in relationships with their friends. It is the time when they begin to ask difficult questions about right and wrong. It is a time of Tom Sawyers and Huckleberry Finns who "promise in blood" and who want more and more and more of life. This is a time when the youngster is particularly sensitive to the kind of life his parents lead. It becomes apparent during the preadolescent years that it is not enough to be a good parent as if that is the only function of being grown-up, but of being a good person, with vast and creative inter-ests in all of life, not merely introverted worry and concern about "*my* child, *my* family, and whether or not *I* am a good parent."

Alan Watts, former Anglican priest and an expert on Eastern philosophy, speaks of the self-defeating attitude of the parent who "sits home worrying about whether you're doing the best thing for your child and living as if the only thing you have to give is a well-brought-up child." He says: "The trouble is that in so many fami-lies the father and mother have been made to feel guilty over whether they are bringing up their children properly. They think the only reason for doing their respective jobs well is to produce a good result in the child. It's like trying to be happy just to be happy. But happiness is a by-product. . . ."*

*A. Watts, "A Redbook Dialogue," *Redbook*, Vol. 127, No. 1 (May 1966).

Just as is a good child. If the only thing the youngster has to look forward to when he becomes a grownup is being a parent who will have to "take care of a brat" (like himself), why bother? Here is where the parents might better ask: What kind of a person am I around my child? rather than, What kind of a parent am I? I want him to be happy. Is there cheer in our home? I want him to be creative. Do I get excited about new things? I want him to learn something. How many books have I read in the past month, year, years? I want him to have friends. How friendly am I? I want him to have ideals. Do I have any? Are they important enough to show in what I do? Have I ever told him what I believe? I want him to be generous. Am I compassionate about the needs of anyone outside my own family? People attract not that which they want but that which they are. People also raise not the children they want but the children who reproduce what they, the parents, are. It is in the "outgoingness" of the parents that the youngsters can begin to see a road that leads away from their own preoccupation with NOT OK. It is "out there," in the world and with people, where the action is, and where, with the stronger and stronger Adult in charge, experiences can take place which begin to produce the OK feelings to counteract the early feelings of NOT OK and despair.

The Adopted Child

The time of preadolescence is particularly difficult for youngsters who struggle with additional burdens. This is the time, for instance, when the adopted child suddenly may break out in bitter rebellion toward his parents in spite of all the well-intended stories he was told about "being chosen." It has long been the standard position of adoption agencies that the little child must be told he is adopted as early as possible, actually long before his Adult is equal to the transaction. All he gets out of it is that he is different. He does not possibly have sufficient data at age three or four to comprehend what adoption means. All he needs to know is that he belongs to someone, to his parents. The finer points of biological birth have insufficient meaning to him at this age. Yet some adoptive parents make such an issue of adoption, of the fact that "we chose you from among all the others," that the little person is left with an obligation he can't possibly pay back. How can I ever be good enough to you when you were so good to choose me? It is the same kind of indignity we see when one person feels the need to say "thank you" to another person for extending the simple courtesy of treating him as a human being—for instance, an elderly person who says "thank you" to a young person for saying "hello" to him. The adopted child's feeling of difference can magnify the NOT OK position until he is a mass of screaming, disorganized frustration. My position in this regard is that discussing adoption should be delayed until the point where the youngster has a sufficiently strong Adult, perhaps at six or seven years of age. Parents may recoil at this and plead the need for "complete honesty with my child." Perhaps a more important principle applies here than abstract honesty, and that is true concern for the little person, who cannot possibly process all the complicated data of this transaction. We step in and protect children from other things they are too young to understand. Why not step in here and protect them from a "truth" they cannot comprehend.

"But he will hear it from the neighbor children!" protest the parents. True, he will. But how this data registers with the little person depends to a large extent on how his parents react. If the four-year-old comes in and reports that the other kids say he is adopted and "what does 'adopted' mean?" Mother can relegate this to something relatively unimportant as she reassures the child that "you belong to us." If the little person is made to feel he truly belongs, be will have a strong enough Adult a little later in life to comprehend why his parents may have postponed telling him the details of his adoption to avoid burdening him with confusing and troubling truth.

We have to examine our absolutes. Is complete disclosure always the best? It might seem to be. However, as Trueblood points out, "We are always guilty of oversimplification when we stress only one of several relevant principles."* He illustrates in the following way that perhaps the concern for the welfare of man or of men is a higher standard and more precious than abstract honesty:

> Consider the consequences of telling the truth in *every* circumstance. Suppose you are in a totalitarian country, in which a man of high principles and courage has been imprisoned. You happen to see him escaping down a certain street and soon afterward you notice that the prison guards are looking for him. You are reasonably certain, that if he is caught and returned to prison, he will be tortured. You are asked whether you saw him go down the street, your only possible answers being Yes or No. What then, in this particular situation, is your moral duty?

Here is a situation in which our decisions clearly must be made by weighing comparative difficulties. This is what parents must do when facing the problem of what to tell the adopted child. It is difficult to tell him he is adopted, and it is difficult not to tell him. Eventually he will know. But the parents can modify the telling in such a way as to protect him from the NOT OK implications by choosing the time, the means, and the details. It is not possible to outline what to say in every situation, for families differ. But it is possible to help the parents recognize the situation of the NOT OK Child and the varying influences of their own P-A-C. With this knowledge these parents can proceed to "play it by ear" and love the child for who he is, *theirs!*

An understanding of the situation of the small child will help parents make those decisions which will produce the maximum stroking, the maximum alleviation of the NOT OK, the maximum support of the truth that "you belong to us." Such understanding also will help adoptive parents to be sensitive to their own NOT OK

*Elton Trueblood, *General Philosophy* (New York: Harper & Row, 1963).

Child. Many people who cannot have children feel so NOT OK that they become excessively demanding of the adopted child: *This child is not going to bring shame on the family,* etc.

The NOT OK burden is even greater for the foster child, but, as with any child, we must start where we are. We cannot go back and reconstruct circumstances into something that did not exist. The usefulness of P-A-C lies in bringing order out of the chaos of feelings, in separating the Parent, Adult, and Child, and in making possible a choice. In my work of many years as consultant to the Child Welfare Division of the Sacramento County Welfare Department, I had the opportunity to work with a great many foster children and their foster parents. I found that if we could develop a sensitivity in both parents and children to the influences of the Parent and Child in both, we could begin to work out the best ways to help these youngsters overcome the powerful, subversive NOT OK recordings made in their early traumatic months and years.

The children of divorce are the orphans of still another storm: the frightening, depressing, emotional storm that tore the family apart. At best divorce is a NOT OK situation, guaranteed to hook the NOT OK Child of all concerned. There is usually very little Adult operating in one of these unfortunate human episodes. This is the major problem. Mother and Father are so totally embattled in crossed transactions that the children are left to muddle through on their own. Even though the parents may be *concerned,* they frequently are not *able* to provide sufficient help to enable the children to live through the family breakup without the fears and humiliations which greatly reinforce the NOT OK. In this situation, as in all situations where children live through periods of great stress, there still is the possibility of their extricating themselves from the tangle of the past if they are helped to recognize that they do have an Adult, which can help them find their own reality and their own way out of the jungle of feelings in which they live.

The Battered Child

The battered child is programmed for homicide. This is the child who has been repeatedly beaten in so brutal a manner that skin and bones are broken.

What is being recorded in the Child and what is being recorded in the Parent of this little person at the time of such a beating?

In the Child are recorded catastrophic feelings of terror, fear, and hatred. The child struggling and thrashing in this nightmare (put yourself in his place) rages inside, If I were as big as you I would kill you! Here is a shift in position to the psychopathic I'M OK—YOU'RE NOT OK. In the Parent is recorded the permission to be cruel, if not to kill, as well as the finer points of how.

In later life this person, under sufficient stress, may give way to these old recordings: He has the desire to kill (Child) and the permission (Parent). And he does!

Many states have enacted battered-child laws, which require that physicians who suspect brutality from the injuries they treat report these suspicions to the authorities. The question is, What happens next? I would say the prognosis is poor unless the child, by the time he is a preadolescent, receives intensive treatment, so he can understand the source of his murderous feelings and further understand that despite his past, he can have a choice about his future. For society to offer less than this to the battered child is to toy with a loaded pistol.

There are, of course, degrees of battering. I believe strongly that all physical abuse of children produces re-playable feelings of violence. The injunction recorded is: When all else fails, *hit!* The final court of appeals is violence. I do not believe in spankings, with one exception: when the child is too little to comprehend danger. A spanking may be the only way to condition him from going into the street. It is most effective in this situation if it is not used daily for nondangerous infractions like spilling the milk or hitting sister. It is not possible to teach nonviolence with violence.

Parents, being more human than not, however, occasionally swing out at their children. The feelings in both parent and child can be talked about with P-A-C so that something constructive may come from the incident: how to keep it from happening again, for instance. It is important for parents to see physical punishment as a take-over by the Child and not as a positive attribute under the heading of discipline.

Bruno Bettelheim says,

> Let's stop for a moment and perform the simple exercise of actually defining the word "discipline." If you go to Webster's, you'll find it has the same root as disciple. Now a disciple isn't somebody you beat over the head. It is somebody who apprentices himself to a master and learns his craft by working at the same vocation. This is the concept of discipline. So if you show your children, "When you're angry you beat; it's a good way to get things," they're going to copy that. And then you complain about violence in our cities.*

Teaching P-A-C to the Retarded

When we recognize that all children struggle under the load of the NOT OK, we begin to appreciate what an oppressive burden is carried by the retarded child. He not only feels NOT OK, he is, in fact, less OK in intellectual endowment than other children. His mental retardation frequently is accompanied by other physical handicaps and visible deformities, which evoke from others responses that underlie his low estimate of himself. In competition with other children, his position is continually reaffirmed, and the acting out of boiling emotions multiplies his problems. He, in fact, has difficulty using the defective computer he has, be-

*B. Bettelheim, "Hypocrisy Breeds the Hippies," *Ladies Home Journal,* March 1968.

cause it is further impaired by the confirming, subversive influence of the NOT OK.

His inability to hold his own in a society of comparison and competition will sometimes create conflicts requiring institutional care, in which this competition is minimized. Yet his emotional turmoil continues to torment him and those around him. The effectiveness of psychotherapy with the retarded is a much debated subject. There is very little in psychiatric literature about treating the retarded. Group treatment has had little trial. The traditional techniques employed in most residential programs include kindly parental control, structured time, avoidance of excessive competition, and an opportunity for relative success with jobs he is able to do. Such techniques have been reasonably successful in providing a secure and sometimes happy life for the retarded. These techniques, however, have been made up mostly of Parent-Child transactions, having but limited effectiveness in helping the youngster to develop inner control by strengthening his Adult. A continual problem for the residential staff has always been the time-consuming business of dealing with emotional episodes.

In Sacramento a new program of teaching P-A-C to the retarded was undertaken in January 1966 by Dennis Marks, a pediatrician, who is the director of Laurel Hills, his recently completed, one-hundred-bed residential center for the retarded. Marks, who is a Member of the Board of Directors of the Institute for Transactional Analysis, came to feel that P-A-C was a system that was so easily understandable that it could be taught to the residents of his center. The age range is from six months to forty-seven years and represents the full range of retardation. Those who attend the P-A-C groups are in the 30-to-75 IQ range. One-third have significant to profound physical handicaps, and many have convulsive disorders. One-third are privately placed, and two-thirds are referred by public agencies such as welfare departments and, occasionally, the probation department. They come from private homes, foster homes, and occasionally state hospitals or juvenile detention centers. In terms of chronological age, most are teen-agers and young adults.

The presence of helplessly handicapped children necessitates the exclusion of children who cannot control aggressive behavior. The open nature of the facility (no locked doors) also requires the exclusion of children who are either extremely destructive or severely antisocial, or those intent on running away. The structure does, however, give exceedingly active and noisy children considerable freedom and acceptance.

The two most urgent problems, therefore, are how to calm the severely agitated, combative child and how to prevent the child from running away. In these two situations, particularly, Marks reports considerable success through the use of Transactional Analysis.

The group of thirty youngsters (we use the term "youngsters" to refer to the entire age range for want of a better generic term) meets once a week in a large living room at the center. They sit in a circle, two deep, from which Marks and the blackboard are visible to everyone. The contract (a term they are comfortable with) is, "We are here to learn P-A-C, which will help us understand how people tick so that we can trade a lot of the uproar for pleasant pastimes and activities." The group is first introduced to the basics of P-A-C: the identification of three parts of a person, represented by the three circles, Parent, Adult, and Child. Marks helps the children identify "what part is talking" when a member of the group makes a statement. For instance, he will ask the group, "Now who's talking?" "Is this John's Parent, or his Adult, or his Child?" In this manner they also learn to identify words. "If you look at a piece of fruit and it's spoiled and you say 'that's *bad*,' that is Adult. If you look at a picture someone is drawing and you don't like it and you say 'that's *bad*,' that is Parent. It is critical and you are making a judgment. If you come running into the playroom in tears and cry 'everyone is *bad* to me,' that is Child. The youngsters very quickly learn to identify words and actions in this way. They find it satisfying and an experience which helps them recognize they have an Adult, or a computer."

The word "computer" is another word with which the children are comfortable. Their understanding of their Adult as a computer has made it possible to talk about retardation, a subject which rarely is mentioned in most institutions. The way Marks presents this to the group is:

> Some guy may have a million-dollar computer and some guy may have a ten-thousand-dollar computer but we're not going to worry about that. All we have to do is find out the best way to use the computer we've got. After all, you don't have to have a million-dollar computer to be nice to people or to do a good job.

Underlying the entire program is the often repeated statement I'M OK—YOU'RE OK. The youngsters repeat this in unison at the beginning and end of each session, and it becomes in their daily living a key that turns off emotions and turns on their Adult. They are helped to understand that comparing is what the Child wants to do. Marks explains:

> The Child wants to say "Mine is better" and "I've got a better computer than you have." That is one way the Child feels better. It is the Child who always is worried about who is smarter. But the Adult can see that, if being smart were the most important thing in life, there would then be only a few happy people in the world: the best painter, or the best mathematician or the best musician; and all the rest would be unhappy because they were not so good. The group grasps and appreciates this approach.

Concerning the problem of controlling aggressive behavior, Marks reports that a severely agitated, combative child can be calmed within two or three minutes. He explains that the groundwork is laid in the group. The methods of restraint are explained as being of three kinds: Parent, Adult, and Child. He has a youngster get up and pretend that he is going to strike him.

"Then I grab his arm and hold it," Marks says, "and ask the group, 'How am I restraining Joe?' " They will agree that this is Adult restraint in that he is merely stopping him from hitting. Then Marks will pretend to hit the child back and they readily identify this as Child restraint. Then Marks will pretend to take him over his knee and give him a spanking, which readily is seen as Parent restraint. The way this understanding is put to use in problems of emotional control was related by Marks as follows:

> I walked into a room one day where there were three people holding one youngster who was in tremendous agitation, trembling with rage, and struggling to hit everyone around him. He was a boy with an IQ of 50 and, on most occasions, was attractive and pleasant. I walked over to him and put my arms around him tightly to restrain him. He was trembling and screaming, "Leave me alone, leave me alone. . . ." After about twenty seconds, I said,
>
> "Now, Tom, how am I restraining you? Is this Parent, Adult, or Child?"
>
> He shouted, "Parent!"
>
> I said, "Not really, Tom. I'm not spanking you. That would be Parent. And I'm not fighting with you. What would that be?"
>
> "That would be Child," he said.
>
> "So how am I restraining you, with my Parent, Adult, or Child?"
>
> "With your Adult," Tom replied.
>
> "OK, that's good, Tom," I said. "Now we'll show these people how we can do it. Now you take my hand and we'll say what we always say." He took my hand, and mumbled, "I'M OK— YOU'RE OK," and we walked together into the TV room, where I suggested he join the youngsters who were in there watching a program.
>
> The whole episode, from encountering a trembling adrenaline-charged, furious child, to walking into the TV room together took exactly three minutes. The key was to turn off the Child and turn on the Adult. This was done by the simple ques-

tion, "How am I restraining you?" There was no way to deal with this angry, boiling mass of feelings called his Child; there was certainly no way at that moment to get to what was bothering him. My objective at the moment was simply to modify his behavior and get past the episode. Nothing "reasonable" could be said or heard while his Child was in control.

The traditional Parent dealing with this situation would have taken considerably longer, with the NOT OK Child suffering more acutely than ever from having been such a "bad boy." As it was, some OK was introduced in the form of mastery by the Adult, the achievement of self-control, and the return to the group activity.

The youngsters easily respond to the imageries of "plugging in the Adult" and turning off the frightened Child or the accusing Parent (as one would a TV set).

Another example Marks gives is one of handling a runaway situation. This is the case of a shy eighteen-year-old girl, with an IQ of 68, who speaks with a tiny voice and usually has very little to say. One day Marks walked by her room and found she was all packed, ready to leave. When she saw him, she blurted, through the tears streaming down her face, "I don't need this place any more. I'm going away!"

The usual parental approach would have been to deny her feelings by something like: "Of course you're not leaving. Now you go to lunch with the other children. You are not going anyplace. And besides, where is your transportation?"

This would only have made her Child more determined and more obstinate and more angry. There is no way to "reason" with the mass of feelings in the Child when the Child is in control.

Marks instead sat down on the girl's bed and said, "You're sure not feeling OK today, Carolyn. Somebody must have really hooked your Child."

"Yes," she responded quickly.

"Well, what happened?" Marks said.

"They won't let me buy a pocketbook," said Carolyn.

"You know," said Marks, "I like your NOT OK Child, but now I want to talk to your Adult. So I'll tell you what . . . you grab my hand and we'll say I'M OK—YOU'RE OK." Which they did. This was the key which had been shaped in the weekly sessions since the beginning of the year. Then Marks was able to talk to her Adult, and her Adult could recognize there was nobody there who could take her shopping that day to get the pocketbook she wanted, and that they perhaps could go tomorrow or the next day. This was simple, once her Adult was back in commission, but impossible as long as the Child was in control. She put her suitcases aside and went into lunch. Elapsed time, four minutes.

"In both of these cases," Marks commented, "we accomplished what we wanted. We subdued the emotional episode and we enriched our relationship. I venture to say that if these youngsters had a sufficient number of these relationships, in a period of a few months, perhaps years, they would learn enough self-control and data processing to enable them to feel and act OK."

Summarily, we may say that the solution to the problems of all children, regardless of their situation, is the same solution that applies to the problems of grownups. We must begin with the realization that we cannot change the past. We must start where we are. We can only separate the past from the present by using the Adult, which can learn to identify the recordings of the Child with its archaic fears and the recordings of the Parent with its disturbing replay of a past reality. Parents who have learned to do this through their understanding and application of P-A-C will find themselves able to help their children differentiate between life as they observed it or were taught it (Parent), life as they felt it (Child), and life as it really is and life as it can be (Adult). They will find that this same procedure will be of the greatest value in the period of change that lies ahead, the years of adolescence, which we examine in the next chapter.

P-A-C and Adolescents

If you wish to converse with me, define your terms.

—Voltaire

One day a sixteen-year-old member of one of my adolescent groups reported the following incident: "I was standing on the street corner and the light was red. My Parent said, 'Don't cross,' my Child said, 'Go ahead anyway,' and while I was debating what to do the light turned green."

The years of adolescence are like this. Teen-agers are confronted with big and little decisions. Yet, often they seem to have to wait for circumstances to make their decisions for them, because they are not really free to decide for themselves. Their brain is nearing its prime development. Their body is mature. But legally and economically they are dependent, and their attempts at emancipated action are frequently undercut by the realization that they can't really make their own decisions anyway, so what's the use of making good decisions. They feel they may as well drift along through adolescence and wait for the light to turn green. The Adult does not develop under these circumstances. Suddenly when they are legally emancipated they feel adrift, they don't know what they want to do, and many of them pass time hoping something will happen, someone will come along, somehow something will turn them on. Yet, at this point, one-fourth of their life has passed.

Because of external and internal pressures the transactions of the teen-ager frequently fall back into the old Child-Parent patterns. In adolescence the feelings of the Child replay in greatly amplified form as the hormones turn on and as the adolescent turns away from his parents as the principal source of stroking to his own age group for stroking of a new kind. The NOT OK tapes come on with increasing frequency, but the coping techniques learned in childhood to minimize the NOT OK now can be dangerous. The seductive cuteness of the little girl must now be brought under control to guard against new developments, both external and internal. The "mine is better" boisterousness of the little boy must be modified in the name of manners as the adolescent learns the painful process of self-control. Communication has to be relearned and revised. The adolescent is pushed out on the stage with a new manuscript in his hands, which he has never read, and the lines don't come off too well at first. He is like a plane shooting ahead at full speed, between converging cloud layers. Below, and rising fast, are the boiling clouds of sexual urges and the rebellious struggling for independence; above are the hovering and lowering clouds of parental anxiety and disapproval. He feels things are closing in, and he desperately looks for an opening.

The central difficulty is that he and his parents often are still working under the terms of the old Parent-Child contract. As much as he sees himself as a grownup, he still *feels* like a child. Parents may suggest what they believe to be a perfectly reasonable course of action and are frustrated, baffled, and hurt over his angry rebuttal, hooking *their* Child. Often the problem is that he mistakes his external parent for his internal Parent. He cannot hear the mother and father of his teen-age years because the old tapes play back the mother and father of the three-year-old, with all the hand slapping, horrified looks, and thunderous "no's" of those early years. The external stimulus hits the Parent, Adult, and Child of the teen-ager simultaneously. The question is, Which one will handle the transaction? Throughout childhood the Child is continually activated, even though there are, depending on the individual, a vast number of Adult transactions. The Child is extremely vulnerable, or "hookable," in this emotionally charged time of life. Whereas the Child responses of the little person could quickly be rationalized as "childish," those same responses now become threatening and disintegrating to the parents. The door slamming of the five-year-old can be rather terrifying if the slammer is a six-foot-tall fifteen-year-old. The sulk of the little girl is seen as ugly and infuriating in the teen-ager. What may have been seen in the little boy as a habit of "making up stories" appears in the adolescent under the heading of lies." The early recordings are the same. Many of the coping techniques of the Child continue in the adolescent years. Bertrand Russell writes of this:

> So many things were forbidden me that I acquired the habit of deceit, in which I persisted up to the age of twenty-one. It became second nature to me to think that whatever I was doing had better be kept to myself, and I have never quite overcome the impulse to hide what I am reading when anybody comes into the room. It is only by a certain effort of the will that I can overcome this impulse.*

This "effort of the will" is the Adult. The Adult can identify the old recordings. It can also recognize the inappropriateness and ineffectiveness of their replay in adolescence. The central need, then, is to keep the Adult in control of this adult-size body so that the realities in the present can gain priority over the realities of the past.

What constitutes the central work of treatment is the freeing

*B. Russell, *The Autobiography of Bertrand Russell* (Boston: Little, Brown, 1967).

up of the Adult in both the teen-ager and his parents in order that an Adult-Adult contract may be drawn. Without an emancipated Adult, life is an unbearable double bind for both. The problem of the adolescent is that inside he has a strong troublemaking Parent, and he is forced to live in the setting in which that Parent developed, where the Parent within is reinforced by the parents without. As parents become threatened and fearful, they find themselves turning more and more frequently to their own Parent for Grandparent solutions, which can be as inadequate as trying to make a jet plane run with hay. Both the parents and the teen-ager are so threatened that the Adult is decommissioned in both. The teen-ager acts out Child feelings and the parents, fearful of letting their feelings take over, most often turn the transaction over to the Parent (grandmother and grandfather). No common reality exists without an Adult-Adult contract, and communication ceases.

I have long admired the Hebrew ceremony of the Bar Mitzvah, which is the symbolic and public drawing of a new contract, or a statement of mutual expectations. In his thirteenth year the Jewish boy becomes a Jewish man, assuming responsibility and religious duty. He does not do so without preparation. This moment has been a goal long-established, and he is prepared for the acceptance of responsibility by rigorous training and discipline as prescribed by Hebrew law. It is unfortunate that a similar event cannot take place in the life of every teen-ager. I know one non-Jewish family who conducted a similar ceremony in their home at the time of their son's fourteenth birthday. He was told that he now was responsible for all his ethical decisions. He accepted this responsibility seriously, although he expressed some concern about the consequences. It will no doubt work well in this case because this young man is prepared for this responsibility. He has been helped to make ethical decisions since early childhood, and he has observed his parents making difficult decisions based in their own binding ethical values.

Frequently teenagers are asked, "What are you going to be?" It is difficult to apply much creative thought to this important question if computer time is continually filled with the unfinished business of "what I have been." Mirra Komarovsky uses the allegory of

> people traveling on a bus in which all the occupants, including the driver, have seats facing the rear. This symbolizes somehow the travels of people through life. But it seems to describe best the travels of the student who at the same time that he is gathering a great storehouse of academia must, in respect to his emotional development, often be looking backward, not forward.*

If this past is understood and filed away, the computer will not continuously be loaded with archaic business but will be free to be creative and get on with the encounters of reality. The teen-ager then can get on a bus where the seats face forward. This way he can have a valid, free choice, can see where he is going and make difficult decisions about where he wants to go rather than accepting fatalistically a route that he didn't choose.

In my practice I have several teen-age groups, which meet weekly. The parents also have opportunities to meet in the evening. The central problem is communication. Repeated crossed transactions have brought conversation to a halt just after "pass the butter" and "I need ten dollars for the weekend." The first step in treatment is to teach both teen-agers and parents the language and concepts of P-A-C. This is an efficient *sorting device,* bringing order out of a mass of chaotic feelings and parental injunctions, which exist in both the teen-ager and his parents. Parents are a mixture of fear, guilt, uncertainty, and wishful thinking. Teen-agers are a mixture of fear, guilt, uncertainty, and wishful thinking. Given a language to explain it, they find they have a great deal in common;

interesting

*M. Komarovsky, "Social Role and the Search for Identity," presented at a symposium on "The Challenge to Women: The Biological Avalanche," University of California Medical School, San Francisco, January 1965.

namely a Parent, Adult, and Child. One of the most neutralizing discoveries for the teen-ager is to find that his parents have a Child, with just as many painful recordings as his own. With this new language the sea of troubles begins to calm. One of my teen-age patients said, "It's really great to be able to talk about ideas at home, and not just people and things." Another said, "The really great thing about P-A-C is that it takes our relationship out of the I-You and breaks it up into six people." In many families the members seem to be prisoners of each other. The youngster says, "You cannot dump parents because you have nowhere else to go." The parent says, "I would love my daughter if she were my neighbor, but I can't stand to live in the same house with her." Through P-A-C this can be talked about as a common predicament, and joint efforts can be turned to not only making the family a bearable group in which to live, but a pleasant and exciting one.

It is not always a simple matter, however, to transform a family from a battleground to a scene of domestic tranquility. Some teen-agers do not easily give up their game of 'It's All Them," even though they may have some insight into its operation. Parents also like to hang on to "Look How Hard I've Tried." When a home situation is particularly uproarious and hostile, an effective way of ringing the bell on the games is to hospitalize the adolescent for a brief period, such as a week. This not only underlines the fact that something is wrong at home, but takes the adolescent out of the Child-provoking home setting and places him in a supportive environment where he can activate his Adult. Then he can begin to learn. At the same time his parents are taught P-A-C and directed to come to the parents' group. When the adolescent is released from the hospital, he joins the out-patient groups for continued treatment.

Unfortunately, treatment sometimes gets off to a bad start by the way in which the teen-ager is brought to treatment. One boy said, "I was pushed into this group and it hooked my NOT OK Child—I didn't know I was coming until the morning of the day

I first came here. We are shoved in here because we are bad; but then you teach us P-A-C and we feel better. However, when we go home we are either made fun of or made uncomfortable. When I try to explain things my old man just cuts it all off with, 'Knock off that P-A-C crap and do as I tell you!' I would really feel better if I could see my parents were equally interested—learning what we are learning. If they only didn't come on in the same old way." The parents of the boy did not attend the parents' group in the beginning. They finally were persuaded to do so and were impressed with how quickly the relationship improved at home.

Some of the most brilliant formulations come from my teenagers. It is as if eight or ten computers are processing data in the group with the aim of extracting new meanings. For instance, in one session a teen-ager said, "I think the Parent is more interested in the institution of the Parent than in the whole individual. Only the Adult can understand that my Child has feelings which are important, too." On another occasion a teen-ager stated: "I think the thinking part of us is a Johnny-come-lately. The feeling part of us was there first. 'I feel' is more encompassing than 'I think.' 'I think' you can back off from, but 'I feel' involves my whole self." Another said, "Only my Adult can honor my father and mother; my Child is too mad."

Many parents are afraid to trust the Adult in their youngsters with hard decisions. One father of a teen-age daughter said, "When she was five and played with a razor I had to take it away from her. Now you see her playing with another kind of razor and what do you say—go ahead and play with it?" The difference is that at five she did not have enough data to comprehend fully the possible fatal consequence of cutting herself with the razor. But at fourteen the teen-ager has, or can have, sufficient data to understand all kinds of consequences—that is, if the parents have been busy through the years acquainting him with values, realities, the importance of people, and his own worth.

Trust in the Adult is the only constructive way to confront the many anxiety-provoking pronouncements the teen-ager can bring home. If the teen-age daughter comes home and announces woefully, "I'm pregnant," it probably will knock the needle off the P-A-C seismograph. The Parent in the parents may rise up in great indignation and judgment; their Child will be tearful and sad (another failure) and angry (how could you do this to us?) and guilty (as the internal Parent whips the Child with its disapproval). What in the parents will meet daughter's announcement? If the Parent and Child stand wringing their hands, one might say the Adult is out boiling water or figuring out what to do. The Adult can determine what part of the Parent and Child can be externalized as constructive data and still contribute something to the daughter's resources for handling this difficult situation. One of the most powerful contributions to inner strength is for the daughter to see her parents struggle with their own desperate feelings and still keep the Adult in control, planning its course on the basis of what is real and what is loving.

She will be needing a great deal of this kind of control herself, in the months ahead. The Adult can process all the realities: the feelings of the parents and of the daughter, the pain of the internal dialogue in both, the extreme NOT OK replaying in both, the shame on the family which both must bear, the difficulty of doing what has to be done, the decision for or against marriage, the decision for or against adoption—in short, the consequences.

In many families an even greater trauma occurs when the daughter announces, "I'm going out tonight with John. He is a Negro." The social stigma on interracial marriages is usually far more severe than that on premarital pregnancy. Some parents have handled this by bellowing, "The hell you are! You let me hear you've been even talking to that boy and I'll bash your silly head in." In that head, of course, is the knowledge that John is class president, comes from a good family, is going to college, and in fact, is quite ideal along with being black. Adding to the dilemma is the fact that in high school she was taught about equality; the class tried to figure out how to end racial prejudice and condemned bigotry. By handling the transaction with the Parent, the parents drive the wedge further into the split between reality and the perception of it. There is another way to handle this—with the Adult, which sees reality not as inimical but as an essential part of the evaluation of what to do. It takes a person of extraordinary perception and integrity to carry out an Adult-Adult biracial relationship. The fact is that society does not yet approve. Neither do the relatives. Neither do most of the church members, despite official pronouncements to the contrary. Some day they may. Does this couple have a strong enough Adult to build a relationship of dignity under these conditions? Some couples have. Can this one? A realistic view of the consequences is the only way to handle this situation. There is a risk, but there is also a possibility of a strengthened Adult, preparing itself for full independence.

One example of the inadequacy of certain parental edicts is in the matter of sex outside of marriage. The red flags of pregnancy and venereal disease, waved more or less successfully by parents through the generations, now both have come down because of the discoveries of science. There is still the very real consequence of bringing shame on the family, although this does not seem as important as it once did, since today nonmarital sexual experience is seen with a positive attitude among many of the parents' peer group. It also is glorified in *Playboy,* in advertising, in the movies, and, in fact, in many aspects of the world of grownups. The Adult view can be quite different as it asks the question, "What does this do to persons?" The Rev. Forest A. Aldrich phrases the predicament in this way:

> Many young people take the attitude that if two people agree to sex and both agree that it is not to be a lasting involvement, and no one is hurt, then what harm is it? The hurt is that something of value—sex—has been devalued. It has been casual and not worth all that could be derived from it. The point is to get through the experience safely. The sin of premarital

sex is not that something was given, but that not enough was given.*

One doctrinal absolute is the evil of using persons as things, even if one of those persons is oneself. If, in the long run, a transitory alliance produces a lack of self-esteem and a reinforcement of the NOT OK position, then sex outside of marriage has provided only a physical release from tension and has not produced the ongoing ecstasy of two people who share unlimited liability for each other. How can one honor this relationship in an unlimited way when there are many others who have a prior claim on one's devotion? Also, many girls report that the experience is unpleasant and they are unable to reach orgasm. "It's supposed to be so great," said one girl. "I don't get it." One boy on being asked if his girl friend reached orgasm said, "Oh, I couldn't ask her that. I didn't know her that well." Sexual intercourse without personal intimacy can only result in a loss of self-esteem. This is true also in marriage.

A splendid book, which examines all of the realities related to sex that confront adolescents, as well as their parents, is Bishop James Pike's *Teen-Agers and Sex*. His central point is that sexual involvement bears with it an ethical responsibility:

> We are dealing not with principles and codes but with the direct effects one's decision may have on other persons, for good or for ill. As Martin Buber, the philosopher-theologian, has so well pointed out, our relationship to God is not I-it, but I-Thou. Therefore any relationship between one human being and another should be I-Thou: a fundamental moral norm that persons are not to be treated as things.†

Whatever the decision in each particular case, the point is that the decision be made responsibly. Bishop Pike continues:

*F. A. Aldrich, Lecture at Fremont Presbyterian Church Family Conference, Zephyr Point, Nevada, Aug. 30, 1966.
†J. Pike, *Teen-Agers and Sex* (Englewood Cliffs, N. J.: Prentice-Hall, 1965).

In the long run, even more fundamental than what precisely our sons and daughters will and will not do is what they understand to be the meaning of the sex act itself—a sacrament, an outward and visible sign of an inward and spiritual grace. The physical act not only expresses the spiritual and emotional involvement of a man and woman; it is also a means whereby that involvement is strengthened. It is a *good* thing. Any restrictions upon it which might be found sound, either from an absolutist or existentialist approach, should be based on the premise that it is a good thing, so good a thing that it should not be utilized under certain circumstances. If restraint is based on the fact that sexual intercourse is so good a thing, rather than on the notion that it is so bad a thing, young people will enter marriage with a much more wholesome attitude, with much greater likelihood of sexual fulfillment in marriage.

The problem remains: How can ideas such as the foregoing be the subject of conversation between an adolescent and a grownup if the two are separated by silence, by awkwardness, by mistrust or exasperation, by the teenager's dogmatic refusal to talk to Mom or Dad ("Why should I? I know the pitch!")? The following conversation between myself and a fifteen-year-old girl illustrates a way in which P-A-C can be used to talk about the complicated relationship problems, including sex, which confront teenagers. At the time of this interview this girl had been seen four times in individual treatment sessions and eight times in treatment groups. This is an individual session. "D" stands for Doctor; "S" stands for Sally (not her real name):

s: You know, you're sounding just like a psychiatrist—of course, you are, but it's just so typical.

d: Is that bad?

s: Well, it's just like a television program that I just despise when they show these psychiatric programs. I hate it. And I am sounding just like a typical patient. I mean, I really am, I know it.

D: Why don't you talk about P-A-C?

S: Oh, I can't today. I can't use it, I am not using it now. I am dealing with everything totally, completely wrong.

D: Do you know what you're saying?

S: No.

D: You're saying to this guy that's acting like a psychiatrist, "I dare you to change me." Isn't that what you're saying?

S: When did I say that to you?

D: Well, that's what you are implying. I ask you, why don't you use your P-A-C, and you say, "I am not using it, I am not going to use it, I dare you to make me use it."

S: I didn't say forever, I just said for today I'm not using it and I don't feel like using it. I'm nervous, that's it. I've been nervous for a couple of days.

D: So you want to play nervous today.

S: No, I don't want to play anything. I want a stronger tranquilizer.

D: You want a stronger *tranquilizer?*

S: Why not? I need a stronger tranquilizer. I shouldn't have come today; you know, I didn't want to come.

D: You want a stronger tranquilizer because you are too lazy to use your P-A-C.

S: I have been using it, and I did try, but I am short-tempered and I—

D: So what's new?

S: So what's new? (laughs) That wasn't nice. But I mean I'm very short-tempered and I wasn't when I got out of the hospital.

D: Is this the only option you have, to be short-tempered?

S: No, it isn't, and I try not to be, but sometimes I am.

D: You blow off the handle?

S: Not extremely, but I find myself getting mad and then when I keep it inside of me it makes me kind of shaky. Do you understand? . . . I hate everything about this and I hate everybody today. I'm going to quit psychiatry. Now doesn't that sound typical patient?

D: With a smile—I'm glad you're smiling.

S: Oh, I can smile about it. I'm . . . it's making me nervous. But do you know what I do?

D: What?

S: If I smile, then I laugh, and then I get very upset, and—

D: Can you tell me what's upsetting you?

S: No!

D: Are you on the verge of tears?

S: I hope not. No, I'm fine. I'm upset today, I knew—I *hate* this. This is getting us nowhere—why don't we just stop my therapy and I'll take pills. What's the matter with me besides headaches and the regular bunch of stuff that seems to be— What is my problem?

D: You don't want to grow.

S: That's what you told me one time. You said I didn't want to grow up. Now that wasn't fair.

D: I don't know what "up" is, I said "grow," you know, open your mind to some new thoughts.

S: To what thoughts?

D: P-A-C.

S: I opened my mind to that when I was in the hospital. I came home and I was feeling pretty good.

D: Why isn't your Adult plugged in today?

S: I don't know.

D: All you can talk about is "I am nervous, I can't, I shouldn't have come here today, you're an old psychiatrist and I'm a patient."

S: Well, that's what we are today.

D: Well, that was a real statement. That came through the Adult. That's what we are today.

S: I can't be P-A-C every single day.

D: Well, it wouldn't be a bad idea. I am.

S: OK, that's fine if you can do it, but I can't do it right now.

D: Oh, why not?

S: Because I'm—

D: Your Child likes to take over.

S: Well, every once in a while I guess maybe it happens. I haven't been using P-A-C all my life or even for a year or anything. I don't know.

D: How are you and your daddy getting along?

S: I have been—I have been very good to my parents.

D: How's your mother?

S: Fine—we have been closer than we have ever been, and I have been affectionate to both of them and I have been trying to be some sort of a daughter that they can like now because, I don't know, I am developing a guilt complex about—I feel that I have been so rotten and everything.

D: Well, let's devote about sixty seconds to that, because I can't see that gets you anywhere—thinking about how rotten you are.

s: If I keep on with this psychiatry I am going to be psychoanalyzing myself from here to doomsday.

D: Is that bad?

s: Well, yes.

D: Not if it turns up some answers.

s: No, it doesn't always. I have a very close friend and he's almost insane, I think. He won't go to a psychiatrist, I have known him for years, and he's so far out of reality it's pathetic, and he psychoanalyzes himself all the time. He reads books.

D: Is he a teen-ager?

s: Yes.

D: Well, it's one thing to psychoanalyze yourself without tools, but you have P-A-C to psychoanalyze yourself, and P-A-C will give you the answers you need.

s: Well, still—OK, I'll tell you something. I don't know whether or not I want to be, to use my Adult all the time. And I try to use it most of the time. Sometimes I just don't want to, it's just kind of a battle, it's almost being actually perfect, it's handling everything the correct way at the right time. It's almost not human sometimes.

D: I know what you mean. Of course we have said before that your Child is what makes you charming and delightful to others, so it isn't that we want to kick the Child out, but let's say that P-A-C is always around, I mean the Parent, the Adult, and the Child are always present. It's true the Child may crowd out the Adult and the emotions take over, or the Parent may crowd out the Adult and the Parent takes over. I guess the trick is to always keep the Adult plugged in even though the Child is playing. If the Child wants to play, let the Adult go along to make sure that everything remains appropriate, because the way that girls get into trouble is when the Child takes over and plays, but plays games that are dangerous. Right?

s: Yes. You mean like being a teaser, something like this?

d: Well—

s: Not knowing when to stop?

d: Yes, right, not being able to— When the Adult is unable to say no to the Child and make it stick, any of us is in trouble.

s: That means for anything, not just for—

d: That's true. Everything. The Child may want to take something that doesn't belong to him or the Child may want to use another person. The Child may want to manipulate another person.

s: Oh, stop talking that way.

d: I have been watching small children manipulate grownups.

s: I manipulate—that's wrong. Right?

d: Well, I don't know whether *wrong* is the appropriate word, but if you manipulate others and it frustrates them or makes them feel bad or upsets them, then I'd say this is something you want to get rid of. Or, if I allow myself to be manipulated, I am going to be upset. If I manipulate others and I don't recognize it but they come at me, then I am upset. See? So where did we learn to manipulate or to allow ourselves to be manipulated? At the age of three, or two.

s: Well, how does, I mean how does it keep on because I was manipulating my father and still do to some degree, I don't know maybe you don't call it manipulation or manipulating people but I could—why, yes, I could. And he allowed himself to be— because I don't know what it was, maybe it was—I manipulated him, maybe I didn't.

D: Well, what goes on between you and your father probably does have elements of manipulation, but part of it is father's enjoyment of his teen-age daughter, you know, he enjoys seeing you happy and enjoys seeing you do things, and he enjoys giving you things, that's part of being a father of a delightful teen-age daughter, but you can take advantage of his generosity, you know, take advantage of his feelings, and this isn't good for you or him either because it leads you into kind of a hassle.

S: I did that.

D: What did you do?

S: I took advantage of him and took advantage of his feelings. I expected to get everything I wanted, I expected every—well, I expected quite a bit and yet he had so much affection for me and I wouldn't let him even touch me unless I was in the mood. I would move away from him and I was actually cruel sometimes. This was even going on in the hospital and then one night I said something horrible when he was bringing me to the hospital, when he was going to go down in the elevator and he wanted to hug me, I think, and I moved away from him and I told him not to do it, and then I just kind of laughed and I said, "Frustrates you, doesn't it?" as if I was really trying to hurt him, and he said, "Yes," and he agreed with me. I felt badly then.

D: Then did you hug him?

S: No.

D: That's a pity, because your Adult would have let your Child hug him because your Adult could have a value that not to hurt anyone is important.

S: I try not to now though, if he wants to hug me I let him. If I don't feel like showing much affection I just let him hug me and that's it. But I have been showing him affection.

D: You don't want to hug him back?

S: Well, now I'll go and kiss him on the cheek or something, like this, and I will show him some affection and I will be very nice and I have done this to my mother, too. I have done this purposefully to show them affection. Well, it wasn't a total act because I mean I felt—

D: You see, the problem here with affection with the opposite sex is the Child is afraid of s-e-x whether it's a feeling or what others may think. Your internal Parent is watching your Child and your Child is fearful of s-e-x because of the internal Parent, but your Adult can say, Look it's perfectly appropriate, perfectly proper for you to give your father affection in the form of a physical embrace and if you can do it, it is a form of mastery of the Adult over the Child.

S: I have been doing it.

D: Good.

S: I have been doing it very well.

D: But you know it's a problem for teen-agers.

S: Well, well, I didn't know that.

D: It really is.

S: Is it?

D: There is a great big word in here called t-a-b-o-o.

S: I don't see why.

D: No? This has existed down through the generations— t-a-b-o-o. It's OK for affection with s-e-x if there are no blood ties. That's a great big taboo. But this is data that needs to be out in the open. I find that I can help any teen-ager to be perfectly natural and affectionate and outgoing with his parents if I can help them to get the data out in the open to process it with their Adult. And you can't be affectionate with the opposite sex, period, you know; it's as if you can't really be discriminating, selective. Once they see it, have the data out in the open, then they are free to be affectionate at the Adult level and affectionate at the Child level and the Adult will take care of the Parent. The Child doesn't have to be afraid of the Parent because the Adult is processing the data with regard to what is real. Parental data is dated, you know. What age?

s: Three.

D: That's right, and that's entirely different from what's real today. And besides, as we both know, you have a handsome daddy and when I see you two together and I see him looking at you I can see that you are the pride and joy of his life.

s: I'm not. I'm so rotten it's pathetic, sometimes.

D: Well, why are you rotten?

s: Because I have given him such misery. I feel sorry. He's a sucker.

D: Well, you probably love him so much that you—you told me once that you have to do these things in order to sort of keep a distance, not get too close.

s: We've always been close, too close, really too close, actually too close.

D: Well, you're his only daughter.

s: Yes, sometimes he, you know—that's unfortunate in some ways too—it's not boys, really I will try. I have a lot of friends who are boys and it's something about them that I don't like, because they do think a lot about sex and usually a lot of them when they look at me, well, they wanted something from me, and—

D: How does that make you feel?

S: Not very good, I don't know, I don't like to be touched, unless I want to be touched and boys do like to touch girls and that bothers me, and I have an awful hard time saying no. I can say no but I just get scared and generally they get the idea, but if one doesn't, I mean I'm going to be cooked, so I've got to watch it.

D: Well, let's see, there are three sets of data always. The Child wants to play, the Parent says "you be ashamed" or "you behave yourself or "you'd better watch out," or you know the Parent comes out with a number of formulas about dealing with a situation like this. The Adult will take into consideration that the Child wants to play, the Parent wants to be disapproving, the Parent has a long list of standards to put into the picture, but the reality part which the Adult is in tune with is: What does the transaction mean to you in reality? For instance, what do you get out of it? What are the dangers? What are the risks? What are the consequences? Do you remember in the group the girls who were in trouble were completely blind to consequences. Of course, we know the Adult is the one who deals with consequences, the Child is not interested in consequences, the Child just wants to play. How many of these teen-agers who were in trouble examined consequences carefully before they made the decision? Well, I'll tell you how many—zero. There are others who have a good Adult. There are a few, I've seen them around. Many of them learned in the group here to develop an Adult.

S: For that, well, it is hard to learn, but those, those are moral values. You get that from your parents; usually. I did. And you learn from each other. Teen-agers do talk among themselves.

D: Well, those are moral values, but they are realistic or Adult values of "Let's don't get hurt" or if you are the most important person in the world, and you should be to yourself, in a way, you don't want to hurt that person and you don't want to get into situations that can louse up, you know, and so on.

S: Do you know what I do?

D: What?

S: I have a tendency to tease, in fact I have been called a teaser before by boys and that's not very good.

D: Well, what do they mean? Do they mean that you are leading them on?

S: Well, sort of, with a gesture or look or just being there or just doing something, sometimes I don't know it and sometimes I do it deliberately.

D: Well, there are two ways to look at this. One is, you are being charming and attractive and delightful to be with, which is good; the other is, you are being seductive which—

S: —is bad and sometimes I even do that.

D: Well, do you know where you learned to do that? Seduction is a game that is learned very early by little girls because it gets them goodies so they are taught early—

S: When?

D: Daddy looks at the little doll and she acts cute and Daddy takes out a piece of candy and takes out a toy and so she is paid off for being cute.

S: (laughing): Maybe that's where I learned. It's my father's fault then.

D: There's no fault there; it's enjoyment for father and daughter.

S: Yes, but you just don't do it to other—

D: Well, it's kind of fun, isn't it?

s: Oh, yes.

d: This is where if the Adult is along with the Child when the Child is playing, like playing seduction, or playing tease or whatever word you want to call it, the Adult will deal with the transaction when the boy makes his pitch—

s: It's not just boys, it's men. If a man looks at me I am flattered, sometimes, if it is not a filthy look, and I actually, sometimes, most of the time, want them to, but yet when they do half of the time or more than half of the time I actually, you know, am not insulted, but get kind of frightened, or not frightened, not usually frightened but I wouldn't look at him twice.

d: What he is saying is and what you are saying: He is saying "Look, I would like to use you," and you're saying, "I know it you so-and-so, but you're not going to get away with it." So here we are back to manipulation, you know. You like to play tease or seduction because it gets you gold stamps. Here we are back to transactional trading stamps again. You say it builds your morale; well, you don't really need this. Every woman plays this game, and it does build the female image, the female morale, but the guy that's giving you the gold stamps wants something in return, you know. And this is what the Adult has to be prepared to deal with so it's— some of these individuals have very attractive Gold Stamps, you know, they've got fourteen-carat gold with lace all around them and it's real difficult you know—big, hand- some, broad shoulders and stuff, but all you have to do is, just like our discussion here, get all the data out in the open and you have a choice once you have processed it through the Adult. You don't have to go all the way, like some girls, because that is their only option. You have a good Adult, you have the option of playing the game up to a point and then saying, Well, it's nice knowing you and then what- ever—

S: Oh, that scares me, I could never let that happen, unless it was by force because it scares me, I don't want anything like that.

D: Why be scared?

S: I don't know but it scares me.

D: Maybe you have to be scared in order to keep your Child from getting out of control, but once you develop confidence in this Adult of yours, and you have a good Adult, you have it made, because your Adult can handle every transaction, even though the Child is enjoying the transaction the Adult still is calling the signals and this is what will save you.

S: I see my time is up. I'll see you when the group starts again. Goodbye.

D: OK, and remember, I'M OK—YOU'RE OK. (End of Interview)

Through the troubling years of adolescence, when young people sometimes seem to turn a deaf ear to the words of their anxious parents, there nevertheless is a hunger to hear and experience reassurances of Mom and Dad's love and concern. The longing for this reassurance was stated in a compelling way recently by my five-year-old daughter, Gretchen. When Mother arrived on the scene, Gretchen was maintaining a precarious balance as she walked along the narrow edge of a brick flower enclosure. Mother said,

"You be careful or you'll fall down into those flowers."

Gretchen said, "Do you care about the flowers or about me?"

The "five-year-old" in the adolescent asks the same question, only he does not state it in so many words. Parents who are sensitive to this unstated plea and who, through acts of love, concern, restraint, and respect, demonstrate repeatedly It Is You We Care About will find the years of adolescence can produce rewards and surprises far beyond their expectations.

11

When Is Treatment Necessary?

We only think when we are confronted with a problem.
—John Dewey

If a person sprains an ankle he can hobble along and eventually the ankle will get well. He still has some use of it while he hobbles. If he breaks a leg he needs support for it while the bone is healing. One problem is an impairment. The other is crippling. Medical attention would have been helpful in the first case; it was imperative in the second.

We can view the need for treatment of emotional problems in somewhat the same way. A person's Adult may be impaired by old recordings from the past, but he may manage to get over difficulties or through problems without treatment. Treatment could make it easier. But he manages. For some people, however, the Adult is impaired to the point where they cannot function. They are crippled by repetitive failure or immobilized by guilt. Frequently there are physical symptoms. Mothers cannot function as mothers, workers cannot do their job, children give up at school, or some persons' behavior becomes inappropriate to the point where they break the law. For these people treatment is required; yet everyone could benefit from it.

All persons can become Transactional Analysts. Treatment simply speeds the process. Treatment with Transactional Analysis is essentially a learning experience through which an individual discovers how to sort out the data that go into his decisions. There is no magic applied by an omnipotent expert. The therapist uses words to convey what he knows and uses in his own transactions to the person who comes into treatment, so he can know and use the same technique. One of my psychiatrist friends said, "One of the best Transactional Analysts I know is a truck driver." The goal is to make every person in treatment an expert in analyzing his own transactions.

Many forms of psychiatric treatment are quite different. The public image is assuredly different. For this reason the decision to go to a psychiatrist generally is not made without a great deal of internal debate. Many patients experience unpleasant feelings over the thought of exposing their inmost problems to someone, even though that someone is an "expert" or professional helper, such as a psychiatrist. As the patient opens the door to the office for the first time, he frequently feels alone, fearful, and ashamed over the implication of failure.

Even if the Adult in the individual gets him to the psychiatrist's office, the Child soon takes over and a Parent-Child situation develops. The patient's Child expresses feelings and anticipates a relationship with the psychiatrist's Parent in the transactions of the first hour. The psychoanalysts refer to this as transference—that is, the situation provokes a transfer of feelings and related behavior from the past, when the patient was a child, into the present, in which the Child in the patient responds as it once did to the authority of the parent. This unique transaction is fairly common in life, and there are elements of it present in any contact with authority, as, for instance, when one is stopped by a highway patrolman. Psychoanalysts maintain that the patient has improved when he has succeeded in avoiding this kind of transfer of feelings from childhood. At this point in analysis, the patient does not have to pick and choose what he is going to reveal about himself to his analyst. In other words, the patient no longer must be afraid of the analyst's Parent. This is referred to in traditional psychoanalysis as overcoming resistance.

In Transactional Analysis we bypass much of the retarding effects of transference and resistance by the mutually participant format and content of P-A-C. The patient soon finds he is relating on equal terms to another human being to whom he has come for help, a human being interested in advancing the patient's knowledge of himself at once so that, as quickly as possible, he can become his own analyst. If the patient is hampered by transference and resistance feelings, these are handled directly with him in the initial hour after he has become acquainted with Parent, Adult, and Child.

In my practice the initial hour has developed into a fairly set pattern in which roughly half of the hour is devoted to hearing the patient's account of his problem and the other half to introducing him to the basics of P-A-C. After the patient understands the meaning of Parent, Adult, and Child, his problem is discussed using the language he has just learned. This transaction "hooks his Adult," to use one of the colloquial expressions Transactional Analysis abounds with, and the patient usually is eager to hear more. The troubled Child doesn't give up easily, however, and may persist or reappear (resistance) in subsequent individual sessions or in the treatment group. An Adult-Adult interpretation is made at each appearance of the Child, pointing out the nature of the transaction originating in the Child and its problem-making burden to the individual's transactions in living.

In the initial phase, Transactional Analysis is essentially a teaching and learning method with the aim of establishing certain specific meanings as a basis for a mutual exploration of how Parent, Adult, and Child appear in today's transactions. This process of establishing a language with specific meanings in the initial phase of treatment is, I believe, unique to this method of treatment and is responsible for expressions of change such as "I feel much better" or "This gives me hope," frequently heard at the end of the first hour.

The initial hour also includes a discussion of the "treatment contract." We use the word "contract" as a statement of mutual expectations (I am here to teach you something and you are here to learn something). It does not imply a guarantee of a cure. It simply states a promise of what the therapist will do and what the patient will do. If either strays from the original expectation, it is a simple matter to review the contract. This dialogue is facilitated by the new language, which opens a way to be specific. The patient agrees to learn the language of Transactional Analysis and to use it in examining his everyday transactions. The goal of treatment is to *cure* the presenting symptom, and the method of treatment is the freeing up of the Adult so that the individual may experience freedom of choice and the creation of new options above and beyond the limiting influences of the past.

Diagnosis

Occasionally during the course of the first hour a patient will ask, "What is my diagnosis?" in a strained voice, braced as if for a pronouncement from on high. This is a provoker for a Parent-Child transaction, which I bypass with a question such as, "Do you need a diagnosis?" or "What would a diagnosis do for you?" It is my belief that more people have been hindered than helped by psychiatric diagnoses. Karl Menninger agrees: "Patients do not come to us to be plastered with a damning index tab. They come to be helped. People can recover from the symptoms of mental illness, but they don't recover from a label."

In the medical tradition diagnosis is an efficient way for physicians to communicate. Knowing the diagnosis helps them to know *what to do*. Acute appendicitis, bursitis, carcinoma of the lung, myocardial infarction—these terms communicate a specific condition and call for a specific treatment. In psychiatric practice the diagnostic tradition has been carried on, but it largely fails in the original purpose of communication. There are many pages of diagnoses in the manual of the American Psychiatric Association, and, with some exceptions, the information each communicates is as vague as the terms Superego, Ego, and Id. To say that a person has a Pseudo-Schizophrenic Obsessive-Compulsive Passive-Dependent Anxiety Neurosis, Chronic does not tell you very much except that this is going to take a long, long time. To say that a person suffers from schizophrenia doesn't really tell you much either, because there is no clear definition of schizophrenia. It may give some comfort to the patient to know that he has such a strange and difficult malady. Few therapists agree on how you treat schizophrenia anyway, or even what the basic unit of observation should be. So diagnostic terms such as these are meaningless and serve mainly to give psychiatric efforts medical respectability and to fulfill the requirements of the hospital records department. Any word that fails in communication is useless and should be discarded. In the final analysis it is what we *know* that makes a difference. Words that obscure the truth must be discarded for those that say it simply, accurately, and directly; and truth about how we are put together, in large measure, is what makes us free.

The language of Transactional Analysis, the observations of an agreed-upon unit (the transaction), and the specific definitions of Parent, Adult, and Child make possible a new, meaningful, colloquial way of communication, not just between doctors who use it but between doctors and patients. A Parent-dominated person with a blocked-out Child knows where his problem lies and can become emancipated from the past without any reference to this being obsessive-compulsive *and* chronic. When a group member insists on knowing his diagnosis ("What am I, anyway?"), I usually respond with a formulation he can understand based in my knowledge of him, which I have gained from observing him in the group. Such a formulation might be as follows: "You have a lot of NOT OK in your Child with a fairly sizable contamination of your Adult, which lets you come on inappropriately at times and gives your hovering Parent an opportunity to beat on your Child. Where do you think all that guilt comes from?"

Dreary preoccupation with symptoms can be just as detrimental as a need for a diagnosis. We have never been able to validate the assumption that a repetitious discussion of symptoms such as depression, headache, insomnia, or abdominal pain did anything for the symptom. We have been able to validate that reducing inner conflict can do wonders for a belly pain. The essential point is that diagnoses and symptoms lend themselves to the rather unfortunate human trait of gaming, or oneupmanship, like "Mine Is Better" or "Nobody Knows the Trouble I've Seen." If a person has problems in his living, regardless of what they are, and he wants help with his problems, he can be taught Transactional Analysis to examine his current transactions in living, as a result of which he can uncover the influences from the remote past that are at the bottom of his troubles.

"How long is this going to take?" is a frequently asked question in the first hour. In many, if not most, psychiatric practices the response to this kind of question has been, at the least, "guarded." The implication has been *a long time*. Jerome D. Frank has pointed out that patients' expectations of the duration of their treatment are primary factors in determining the length of time it takes to achieve similar therapeutic results. He cites two groups of similar "psychosomatic" patients who arrived at the same end point, one in six weeks and the other in one year, depending upon their anticipations regarding the length of time that would be needed. I think the key to anticipation is an understanding of therapeutic results to be achieved.

Our goal in treatment is clearly stated with the new language just learned, and thus the patient knows what he is undertaking. I like to help my patients think of limitations imposed by the realities of time and cost as a challenge rather than a depriving imposition. This often is stated in the proposal, "Let's schedule you to come to the two o'clock Tuesday group for ten sessions and see how much we can do in that time." If the patient wants to continue after this series, we can set up another ten sessions. He knows he can come back. The average length of time in a group in my practice is twenty hours. There are variables, of course, beginning with individual differences. We vary in Parent, Adult, and Child. We vary in difficulties in our living situation: marital problems, unsatisfying work, no leisure-time outlets, etc. There have been patients who have achieved a breakthrough in three or four group sessions; that is, they achieved enough freeing up of their Adult that they could begin to accurately differentiate their Parent from the Child and both from reality—the outside world.

One of the first indications of this differentiation is the statement by the patient, "My NOT OK Child was . . ." or "is. . . ." The use of this expression signals the achievement of an understood, authentic, and real separation of the patient's Child from his Adult—that is, it is both intellectually and internally and externally integrated into his personality.

Why Group Treatment?

The treatment of individuals in groups is the method of choice by Transactional Analysts. Is this good or bad? Is the treatment of individuals in a group "bargain-basement psychiatry"? Many people react to the word "group" as they reacted to Franklin Roosevelt's term "the common man." Who wants to be common? Who wants to be depersonalized into a statistic or into mere membership in a group? What goes on in group treatment? What goes on in group treatment using Transactional Analysis?

One common impression is that in group treatment people come to express feelings, "get it out of their system," tell other people what they think of them, and "anything goes." In fact, many writings about group treatment have encouraged this point of view. S. R. Slavson, one of the pioneers in the development of group treatment methods, stated in his book *The Practice of Group Therapy:*

> The chief and common value of the group is that it permits acting out of instinctual drives, which is accelerated by the catalytic effect of the other members. There is less caution and greater abandon in a group where the members find support in one another and the fear of self-revelation is strikingly reduced. As a result, patients reveal their problems more easily, and therapy is speeded up. Defenses are diminished, the permissiveness of the total environment and the example set by others allow each to let go with decreased self-protective restraint. Although groups lessen the defenses of adults as well, this is particularly true of children and adolescents. Free acting out and talking through yields satisfaction. At the same time it brings patients face to face with their problems quite early in treatment. The defenses against injury to one's self-esteem are also reduced. The friendly group climate and the mutual acceptance do not require one to be on the defensive. All have the same or similar problems and no negative reactions are anticipated by anyone. Status is assured. There is no fear of retaliation or debasement.*

In my own clinical experience, *I have not been able to validate the above statement.* Allowing the Child to come on, act out instinctual impulses, and play games at random in the treatment group is a waste of the group's time and invasion of the rights and purposes of each individual group member. If permitted to continue, it sabotages the therapeutic contract of Transactional Analysis. Until each member of the group has achieved at least some

*S. R. Slavson, *The Practice of Group Therapy* (New York: International Universities Press, 1947).

freeing-up of his Adult, self-revelation, or the confessional, contributes very little, if anything, to the purpose of curing individuals in the group. Treatment is speeded only by keeping the Adult in charge. Only the Adult can spot the Child or the Parent. Revealing problems is an invitation to play, "Why Don't You, Yes But." Expressing feelings and "talking through" may yield satisfactions to the Parent and the Child, as in everyday life, but in the treatment group such transactions interfere with the acquisition of basic understandings and concepts essential to achieving an emancipated Adult.

There is no magic in the word "group." Since in its initial phase Transactional Analysis is a teaching-learning experience, the group setting has several distinct advantages over the traditional one-to-one setting of individual treatment. Everything that is said in the group needs to be seen and heard by every other member of the group—every question, every answer, every transaction. The subtle and multiple ways the Parent reveals itself in transactions need to be identified and learned. Both the inner and outer threats to the Child need to be recognized in a general sense at first, then the unique and specific characteristics of the Child in each individual in the group. There is a mutual confrontation of games, of realities "where you live," which is quite different from the essentially seclusive and permissive "ear" in one-to-one therapies. In group treatment people are seen in the natural milieu, involved with other people, instead of all by themselves in a separateness that can never be duplicated on the outside. The primary benefit of treating people in groups with Transactional Analysis is that they move faster toward getting well, starting to live, beginning to see and feel what is real, or "growing up," however the individual's goal in treatment might be expressed. At the end of an exciting hour in the group, one member said, "I feel like I'm ten feet tall."

Before this primary benefit is examined, however, it may be well to point to the answer group treatment provides to the widely acknowledged high cost of individual treatment and the disparity between the number of people who need help and the number of people available to give it. We live in an age of cost and time consciousness. We are surrounded by a sense of urgency to help people in trouble, many people in trouble. In finding a solution we must examine one of the principal criticisms of psychiatric treatment: It costs too much; it takes too long for the uncertain results achieved. We cannot dismiss this criticism simply by countering with the judgment that people who hold this view do not have realistic priorities, as, for example, the individual in real trouble who sees his solution in owning a new model car rather than seeking much-needed help.

There are many people today who, although they may wholly accept the idea that "mental health is important," still have not the means to add the burden of long-term psychiatric treatment to their already consumptive costs of living. In this category fall many of the middle class and all those in low-income groups. Is mental health for the wealthy? Is psychiatric treatment, as I heard a physician colleague define it recently, "a luxury"? Or can many more people be helped by group treatment? Can psychiatric care be considered as realistic a part of the treatment as emergency surgery?

Dr. Leonard Schatzman, a medical sociologist at the University of California Medical Center, in 1966 completed a field study of fifteen medical centers over a period of eight years, concentrating on psychiatrists and their staffs. He observed in an article in 1966, published in the *San Francisco Chronicle:*

> The older, one-to-one medical treatment model for the wealthy, and the efficient neglect of the masses of the poor, are no longer seen as adequate. Upgraded populations today are demanding more psychiatric goods and services. The psychoanalytically oriented psychiatrist sticks close to his office, provides personal service to a very limited clientele and has to deal with wealthier people in order to cover the cost of his operations. Good, bad, or indifferent, the service is tailored to the client, and with great pride and finesse. But who is buying tailor-made suits? Who is

regularly dining in gourmet restaurants with candlelight and wine? Who drives custom-made cars?

The treatment of individuals in groups can reduce treatment costs to the point where most wage earners can afford it. It has been my experience, also, that group treatment using Transactional Analysis has reduced the duration of treatment, which also has resulted in a reduction in cost to the patient. A third factor is that the "contract" for treatment, and the procedures used, are so specific that I believe this treatment format would be insurable on a large scale. If we can buy insurance to provide our children education, it would seem that we can insure a special kind of education about behavior also.

More important than these considerations, however, is the fact that, in my experience, individuals get well faster in group treatment using Transactional Analysis than in the traditional one-to-one treatment relationship. By "get well" I mean achieving the goals stated in the initial hour contract, one of which is the alleviation of the presenting symptom (e.g., marriage breaking up, fatigue, headaches, job failure, etc.) and the other of which is to learn to use P-A-C accurately and effectively. One measurement of the patient's cure is whether or not he can report what happened in any transaction in a way understandable to others in the group. If someone tells me he was in therapy for a long period of time and "it was very helpful," yet cannot answer my question, "what happened in therapy?" then I do not feel he has achieved mastery over his own actions. Applicable here is the Aristotelian idea that "that which is expressed is impressed." If a patient can put into words why he does what he does and how he has stopped doing it, then he is cured, in that he knows what the cure is and can use it again and again.

Once a patient has learned the fundamentals of P-A-C, he can see the Transactional Analysis group as something much different from that which he is conditioned to see by his Parent and Child.

He may have been taught early not to "wash your dirty linen in front of others" or "don't give out the family secrets." This comes through as a clearly recorded tape from the Parent. The Child on the other hand "wants the floor the whole hour" in an ongoing game of "Poor Me." An individual who wants to play "Confession," "Psychiatry," "Ain't It Awful," and "It's All Him" soon finds no one in the group cares to play with him. The role of the therapist is that of a teacher, trainer, and resource person with heavy emphasis on involvement. The group is the setting of activity, involvement, and movement with sustained permission for laughter to provide a relaxing release from any tendency to see the experience as "grim business indeed."

The goal for each member of the P-A-C group is clear, concise, and easily stated: to *cure* the patient by freeing-up his Adult from the troublemaking influences and demands of his Parent and Child. The goal is achieved by teaching each member of the group how to recognize, identify, and describe the Parent, Adult, and Child as each appears in transactions in the group.

Since the essential characteristic of the group is that of teaching, learning, and analyzing, the effectiveness of the Transactional Analyst rests in his enthusiasm and ability as a teacher and his alertness in keeping abreast of every communication or signal in the group, verbal or otherwise. In the setting of the group the Parent appears in a multitude of ways: the wagging index finger, raised eyebrows, pursed lips, or statements such as: "Don't you agree?" "*Everybody* knows that . . ." "*They* say . . ." "After all . . ." "I'm going to get to the bottom of this, *once and for all!*"

The Child also makes its appearance in easily recognized ways: crying, laughing, coyness, nail biting, fidgeting, withdrawal, and sulking, in addition to the variety of Child games such as "Poor Me," "Ain't It Awful," and "There I Go Again." The members of the group are supportive of the NOT OK Child in another member and seldom, in Parent fashion, *accuse* his Child of coming on. Instead there is a sympathetic approach such as, "I can see your

Child is hurt; how come?" or "Can you tell me what hooked your Child?"

Through multitudes of transactions in the group the patients quickly begin to fill in the gaps of information about each other's Parent, Adult, and Child. This is a "team evaluation," not of data long since buried, but of observable data that manifests itself in the open, today, in transactions that involve each other. The team is made up of participants, however, and not of antiseptic observers called a treatment team. Few patients will tolerate the team approach, and few psychiatrists can justify it to their patients, says Avrohm Jacobson, Director of Psychiatry at the Jersey Shore Medical Center:

> Clinics, however, continue to "evaluate" patients via the team approach. This is a cruelly lengthy process for the patient, involving a ritual of data-gathering by the caseworker [Archaeology] and testing by the psychologist that contributes very little to the psychiatrist's clinical impression. . . . The time he must spend in conference listening to all the reports—so carefully gathered over a period of several months—could be more usefully spent in direct contact with the patient.

He referred to an earlier study of one clinic, which demonstrated that most of the clinic time was devoted to the work-up of patients who would not be seen in therapy.*

In the early days of my use of P-A-C some patients were wary of entering a group, insisting, according to their understanding of traditional treatment methods, that a private and repetitious recital of problems was what they had come for. Their position was: I am paying you to listen to me and somehow something will come of it. This attitude has largely changed because of the good reports of

---

*A. Jacobson, "A Critical Look at the Community Psychiatric Clinic," *Community Psychiatry,* Supplement to *The American Journal of Psychiatry,* Vol. 124, No. 4 (October 1967).

the effectiveness of group treatment. More recently they are referred directly to the group from outside sources, or they ask to be allowed to enter a group, having heard about such a group from a friend. There is no selection of members for the group according to diagnostic categories. Nor are they assigned to groups on the basis of symptomatic similarity, not only because it isn't necessary but because of the stigmatizing aspects of psychiatric diagnoses. It is not beneficial to put all alcoholics, all homosexuals, or all school dropouts in the same group, since this makes possible the development of the general tenor "Doesn't everybody?" with the therapist the only odd one.

Thus the group may include all standard diagnostic categories, including patients with low intelligence and those lacking in formal schooling. Many "self-taught" individuals make good Transactional Analysts. Many of my patients have had the opportunity to see a patient in the group go into and come out of an acute psychotic episode (decommissioned Adult) and the free expression of numerous delusions (take-over by the archaic Child). In the group they have observed and heard patients who were actively hallucinating describe the Parent-Child dialogue that the patient perceived as coming from outside of himself. Patients with freed-up Adults are not disturbed by these manifestations of transitory mental disturbance. They tend to be supportive, reassuring, and stroking, and to ignore the unusual.

Each of my Transactional Analysis groups meets weekly except for hospital groups, which meet daily. At the end of the hospital stay, the duration of which averages two weeks, the patient enters one of the groups in my office. Group members are taught to be alert to the tendency of the Child to compare—"I am learning faster than you," or, "You are sicker than I." Therefore, new patients entering a group of "old-timers" seem to feel at ease, and quickly proceed to the business of Transactional Analysis. The setting for group sessions is comfortable and acoustically perfect. Everything can be heard, including a sigh. Occupying a prominent place in the room is a blackboard, which is used frequently in each session for structural diagrams of the symbolic rendering of important formulations.

Some people move rapidly in the acquisition of skill in identifying Parent, Adult, and Child and the ways these are involved in current transactions. Others take longer. Yet, those for whom learning may come more slowly develop the insight, in time, that their resistance to learning resides in the NOT OK Child that is laboring under an old reality in which the little person was not given permission to think for himself.

The understanding of the existence in oneself of the NOT OK Child is one of the first and most important steps in understanding the basis of behavior. This marks the beginning of the objective evaluation of one's own personality structure. It is one thing to understand this academically. It is another to comprehend this reality in oneself. *The* NOT OK Child may be perceived as an interesting idea. *My* NOT OK Child is real.

The content of group transactions is related mostly to the present-day problems of the members. What happened yesterday, or what happened last week, is much more often the point of inquiry than what happened a long time ago. The members learn to identify and know their Parent, Adult, and Child by their appearance in the transactions of the present, particularly the transaction in the group itself. This is quite different from the kinds of data we sometimes think of as coming from psychological research. In an address to the American Psychological Association in September, 1967, the Association's president; Abraham Maslow, asserted his colleagues generally are far too fond of amassing "trivial" facts under the banner of research. "The information they gather is useful, but it tends to be trivial, tends to be a piling up of little facts. . . . Far too many psychologists do their work on refined subjects such as 'the left quadrant of somebody's eyeball.' "*

*San Francisco *Chronicle,* Sept. 15, 1967.

The ultimate value of research, whatever its form, lies in the production of information that enables people to change. The change produced in individuals as their Adult begins to take charge is readily apparent in the group. It also becomes apparent to other members of the family. Not infrequently this may present certain hazards for the individual. A husband whose wife was in one of my groups called to complain: "What gives in that group— my wife seems happier, but our marriage is going on the rocks." In such a case I invite the spouse in for an individual session to explain the basics of P-A-C. The usual outcome is that both husband and wife enter a marriage group. It is almost axiomatic that if one member of a family enters a group and begins to change, the whole family must become involved, because the game pattern has been disturbed.

If, for instance, one member of a family is "the black sheep" and he begins to move out of this role, the roles of others, particularly siblings, may become confused, reversed, or otherwise upset. This is the basis for the usually excellent results achieved with conjoint family therapy. In my adolescent groups, the contract calls for equal involvement of the parents. One of the repeated topics for discussion at these group meetings is "How to Sabotage Therapy." Some parents unknowingly undermine treatment efforts because they really do not want to give up the Parent-Child relationship which they feel has "worked so well" in the past. Their position of power is threatened when the adolescent starts operating in the Adult, and unless the parents come on Adult, the transactions are crossed. These parents see autonomy in their youngster as a threat to their control of him and may decide they liked it better the way it was, before treatment. Familiar miseries may seem more comfortable to frightened parents than the risk of trusting their teenager to develop his own inner controls.

Group members are encouraged to view their relationships on the outside in a responsible and loving way. Some relationships exist by virtue of games. To stop playing is to end the relationship. This is not always kind or realistic. If visits to Grandmother's house have been structured for the past twenty years by games of the "Ain't It Awful" variety, it is not necessarily loving to stop visiting Grandmother because you can no longer stand "Ain't It Awful." The Adult has a choice: to play, to not play, to modify the game into something less destructive, or to try to explain the insights that help persons give up games. We cannot, after all, resign from the human race, game-ridden as it may be. If we are not to be overcome by evil, then we must overcome evil with good. This we cannot do if we withdraw from all the relationships in which games exist.

From time to time I refer to the built-in safeguards in P-A-C. As I write this I am confronted with rows of book shelves loaded with tomes devoted to the general topic of therapy. Much of the content is devoted to repetitious, morbid accounts of so-called "mental illness" or human misery, with minutely detailed "technical" discussions of the dangers involved in therapy. Much of this has to do with so-called transference and resistance problems so central to the method of psychoanalysis. Too often these writings dwell on how to protect the therapist rather than how to cure the patient. In psychoanalysis the analyst is the hero. In Transactional Analysis the patient is the hero. The safeguards in P-A-C exist in its mutually participant format with a language that forms the basis for patient-to-patient and patient-to-therapist transactions for the meaningful examination of all aspects of behavior and feelings regardless of their nature. In the P-A-C group the members act as both a restraining and supportive influence to each other. There is nothing of the omnipotent therapist sitting in the dark corner with his poor little patient recumbent before him, both alert to the dangers in the grim business. One aspect of the P-A-C group contract allows and even encourages the Child in each member, including the therapist, to come out and laugh. P-A-C groups are characteristically laughing groups with great capacity in turn to be considerate and supportive with the nurturing Parent while looking for new answers with the alert Adult.

The danger, then, is the therapist's not knowing or, for that matter, anyone's not knowing what the I'M NOT OK position in the Child can do to a person's own life and to the lives of others around him. When one member in the group announces, "You hooked my NOT OK Child when you said that," the way is open for the examination of one of the mysteries of our existence, the outcome of which may prove to be exceedingly beneficial to all the members of the group.

12

P-A-C and Moral Values

> I submit that the tension between science and faith
> should be resolved not in terms either of elimination
> or duality, but in terms of a synthesis.
>
> **—Teilhard de Chardin**

You tell your six-year-old son to go back out there and punch that kid in the nose the "way he punched you!" Why?

You march in a demonstration protesting the Vietnam war. Why?

You give one-tenth of your income to your church. Why?

You do not report your good friend to the Internal Revenue Service, although you know he is guilty of gross tax evasion. Why?

You accept responsibility for the mistake of an employee. Why?

You are for fair-housing laws but forget to vote. Why?

You tell your daughter she must stop associating with a certain friend who comes from an undesirable home. Why?

You do not report the malpractice of a colleague even though you know people are being harmed. Why?

You do not allow your children to watch *Divorce Court,* but they may watch *The Man from U.N.C.L.E.* Why?

Every day most people make decisions of this kind. They are all moral decisions, or decisions of right and wrong. Where does the data which goes into these decisions come from? From the Parent, Adult, and Child. After you have examined all of your own Parent data, kept some and rejected some, what do you do if you do not feel you have the necessary guides for decision making? Abdicate? Once you have an emancipated Adult, what do you do with it? On moral questions, can you figure things out for yourself—or do you have to go ask an "authority"? Can we all be moralists? Or is that for very smart and wise people?

If we don't seem to be doing very well, where can we go for new data? Where are we deficient? What kinds of reality can the Adult examine?

Reality is our most important treatment tool. Reality, understood through the study of history and the observation of man, is also the tool by which we construct a valid ethical system. We are not reasonable, however, if we assume that the only reality about man is that within our own personal experience or comprehension. Reality, for some people, is broader than it is for others, because they have looked more, lived more, read more, experienced more, and thought more. Or their reality simply is different from someone else's reality.

Our need for direction in the journey through life is similar to the navigational problem of an airplane pilot. Pilots in the early days of aviation flew "by the seat of their pants" and relied on their vision, comparing what they saw below them—rivers, inlets, railroad tracks, and towns—with the maps they had spread before them. This, of course, was unreliable when vision was obstructed, even for a short time. Therefore, navigational aids were devised to "take a fix" on two points. (The two points are special radio stations. Each emits a signal informing him of the compass radial his plane is on in relation to the station.) He draws the two radials as lines on his map, and where the two lines cross is where he is. If he took a fix in only one direction he could not find his location. He might discover he was on the equator. But *where* on the equator? He would have to "look" in another direction for the data to answer that question.

I feel that many psychiatrists and psychologists have been guilty of "one fix" treatment in that they have devoted all their time to looking at only one reality, the past history of the patient—*what he did*—and largely ignored an examination of the types of reality that might help him understand *what he should do*.

We are hopelessly impoverished if we believe that the only realities that concern our mental health have to do with a state of affairs wherein "I am So and So because when I was three years old Mother hit Father with my potty seat on Christmas Eve in Cincinnati." Archaeology of this kind reminds me of H. Allen Smith's story about the little girl who wrote a thank-you note to her grandmother for giving her a book about penguins for Christmas: "Dear Grandmother, Thank you very much for the nice book you sent me for Christmas. This book gives me more information about penguins than I care to have."

We can spend a lifetime digging through the bones of past experience, as if this were the only place reality existed, and completely ignore other compelling realities. *One such reality is the need for and existence of a system of moral values.*

Establishing value judgments has been seen by many "psychological scientists" as an abominable departure from the scientific method, to be shunned righteously, and at all cost. Some of these people steadfastly insist that scientific inquiry cannot be applied to this field. "*That* is a value judgment; therefore, we cannot examine it." "*That* is in the field of beliefs; therefore, we cannot assemble plausible data." What they overlook is the fact that the scientific method itself is totally dependent on a moral value—the trustworthiness of the reporters of scientific observation. Why does a scientist tell the truth? Because he can prove in a laboratory that he should? Nathaniel Branden has devoted a paper to the serious problem raised by those who hold that scientists have no business concerning themselves with moral values:

> Central to the science of psychology is the issue or problem of *motivation*. The base of the science is the need to answer two fundamental questions: Why does a man act as he does? What would be required for a man to act differently? The key to motivation lies in the realm of values. The tragedy of psychology today is that *values* is the one issue specifically banned from its domain. It is not true that merely bringing conflicts into conscious awareness guarantees that patients will resolve them. The answers to moral problems are not self-evident; they require a process of complex philosophical thought and analysis. Effective psychotherapy requires a conscious, rational, scientific code of ethics—a system of values based on the facts of reality and geared to the needs of man's life on earth.*

Branden charges that psychiatrists and psychologists bear a grave moral responsibility if they declare that "philosophical and moral issues do not concern them, that science cannot pronounce value judgments," if they "shrug off their professional obligations with the assertion that a rational code of morality is impossible and, by their silence, lend their sanction to spiritual murder."

What Is a Rational Code of Morality?

A frequent response to such a question as this is, "If everyone lived by the Golden Rule everything would be fine." The inadequacy of this answer lies in the fact that what we do unto others, even

*Nathaniel Branden, "Psychotherapy and the Objectivist Ethics," delivered before the Psychiatric Division of the San Mateo County Medical Society, Jan. 24, 1966.

though it may be what we would have them do unto us, may be destructive. A person who tries to solve his NOT OK by a hard and continuous game of "Kick Me" does no one a favor by projecting this "solution" on someone else. The Golden Rule is not an adequate guide, not because the ideal is wrong, but because most people do not have enough data about what they want for themselves, or why they want it. They do not recognize the I'M NOT OK—YOU'RE OK position and are unaware of the games they play to relieve the burden. Persons stop taking the Golden Rule and many similar "beliefs" seriously, because in their own experience these beliefs do not work.

Bertrand Russell writes:

> Many adults in their hearts still believe all that they were taught in childhood and feel wicked when their lives do not conform to the maxims of the Sunday School. The harm done is not merely to introduce a division between the conscious reasonable personality [Adult] and the unconscious infantile personality [Child]; the harm lies also in the fact that the valid parts of conventional morality become discredited along with the invalid parts. This danger is inseparable from a system which teaches the young, *en bloc,* a number of beliefs that they are almost sure to discard when they become mature.*

Are there then, as Russell suggests, "valid parts of conventional morality"? One function of the freed-up Adult is to examine the Parent so that it may have a choice of accepting or rejecting Parent data. We must guard against the dogma of rejecting the Parent *in toto,* and ask, Is there anything left worth saving? It is clear that much Parent data is reliable. It is, after all, through the Parent that our culture has been transmitted.

Moral values thus can be seen as appearing first in the Parent. We think of "should" and "ought" as Parent words. *The central question of this chapter is: Can "should" and "ought" be Adult words?*

*B. Russell, *Why I Am Not a Christian* (New York: Simon & Schuster, 1957).

Is Agreement on Moral Values Possible?

Is there an objective morality that has claims on all men, or must we construct our own individual, situational moralities? Viktor Frankl comments on the despair of youth today who find themselves in what he calls an existential vacuum, where each person is the center of his own universe, where there is a denial that there are any claims upon him which come from "without" himself.* All morality in this vacuum is subjective. If this is true, we must then consider that in the world today there are three billion "moralities," with three billion people going their own way, denying that any objective principles may govern the relatedness between people. Yet the fact is that, the search for these objective principles and the longing for relatedness is a universal reality. It is also felt as a personal, experiential reality. The fact is that people cannot and do not want to live unrelated to other people. Some persons who are devoted to the use of LSD base their devotion on what they report as the transcendence of the psychedelic experience, that "out there" they are discovering a common essence that ties all men together. Though their vehicle of transcendence may be controversial, we must take into account this longing for relatedness, the capacity for the feeling of oneness, and the conviction that human beings, because they are related to one another, have claims on one another.

The longing for relatedness is a fact, even though the principles governing that relatedness cannot be arrived at empirically. The strongest arguments for ethical objectivity, states Trueblood, are not empirical but always dialectical. He states,

> When we face these carefully, noting that subjective relativism can be reduced to absurdity, we are driven to believe in an existent moral order, even though our understanding of it in any one period or in any one culture may be dim indeed. *What then do*

*V. Frankl, address, Sacramento State College, May 5, 1966.

discussing/
truth
opinions; debate
reasoning

we mean by an objective moral order? We mean that reality in reference to which a person is wrong when he makes a false moral choice, either in his own conduct or in the judgment of another. The conclusion that there is such an order, which the (dialectic) requires, is not the same as knowing precisely or even approximately what the nature of the moral requirement is. When men differ about moral standards, it does not mean that they should give up the struggle to learn what they ought to do.* [Italics mine.]

Those who reject the idea that there is an objective moral order, or a universal "should," must stop to consider the difficulties implied by this rejection. The existentialists have rejected this notion. Sartre contended that man creates his own human essence through a series of choices, of acts that fashion him. He maintained that man, through his actions, creates his own definition of man, that man's existence, in short, precedes his human essence. Not only does man create his own essential humanity, but he simultaneously creates all human dignity. He can only choose what is good for him, but what is good for him must be good for all men.

Joseph Collignon reminds us, however, that there is a reverse of the coin:

Man must therefore take the responsibility for every act, not only for himself but for all men. It is not without reason, then, that Sartre finds "anguish, abandonment and despair" a part of his lot—and the lot of every existentialist. For if no one and no creed can help one in a decision, *cosmic in its significance,* one can readily imagine the despair implicit in such a philosophy. . . . Existentialism has a strong appeal for the young. There is a thrill in thinking the world absurd, for it gives them a sense of superiority over the established order, of mastery over them-

---

*Elton Trueblood, *General Philosophy* (New York: Harper, 1963).

selves. The world for them ceases to have cut-and-dried philosophical unity; there is room for action, in creating, if only for themselves, a human dignity.

But disillusionment is there, too. At the completion of a lecture on existentialism about a year ago, I found many of the students enthusiastically taking the philosophy to heart. The final lecture was interrupted by the crushing news of President Kennedy's death. In the stunned silence that followed, one voice jarred, brash and strident: "It was a perfect existential act." Though he was quieted in no uncertain terms by the class, many of whom were crying, the thought persisted: Yes, it *was* a perfect existential act. Nothing had to be said; to act individually, freely was wonderful, but who would control the free action of assassinating a young President who had given most of his adult years to devoted service of his country? The act of killing a President may have been a rewarding experience in the free exercise of will for Lee Oswald, but for the rest of the country and the world . . .* [Italics mine.]

If there is no universal "should," there is no way of saying that Albert Schweitzer was a better man than Adolf Hitler. If not, the only valid observation we may make is that Albert Schweitzer did such and so, and Adolf Hitler did such and so. Even though we make further notations that Albert Schweitzer saved so and so many lives and Adolf Hitler instigated the death of millions of people, we see this only as statistical markings on the page of history and discount any relevance of ethical reflection toward the modification of human behavior. The worth of people, or persons, after all, cannot be proven scientifically. Albert Schweitzer thought he was right. Adolf Hitler thought he was right. That they were both right is an obvious contradiction. *But by what standard do we determine who was right?*

*J. Collignon, "The Uses of Guilt," *Saturday Review of Literature,* Oct. 31, 1964.

The Worth of Persons

I would like to suggest that a reasonable approximation of this objective moral order, or of ultimate truth, is that *persons are important* in that they are all bound together in a universal relatedness which transcends their own personal existence. Is this a reasonable postulate? The most helpful analytic concept in attempting to answer this question is the concept of *comparative difficulties*. It is difficult to believe that persons are important, and it is also difficult to believe they are not.

The denial of the importance of persons negates all our efforts in their behalf. Why all this elaborate fuss about psychiatry if persons are not important? The idea that persons are important is a *moral idea* without which any system of understanding man is futile. Yet we cannot establish this importance at the end of a syllogism. History, ancient and modern, with its detailed account of denigration and human destruction, would seem, conversely, to substantiate a position that human beings are of little consequence. The birth, torment, and death of the billions of persons who have lived on this earth, if there were no direction or design to human existence, would seem more logically to support a position of futility in all our efforts to understand men's minds and to bring changes in human behavior. We cannot prove they are important. We have only the faith to believe they are, because of the greater difficulty of believing they are not.

"A man will continue to research," wrote Teilhard de Chardin, "only so long as he is prompted by some passionate interest, and this interest will be dependent on the conviction, strictly undemonstrable to science, that the universe has a direction."*

We are not honest men of science if we disregard the fact that this "passionate interest" has, in fact, persisted throughout the his-

*Pierre Teilhard de Chardin, *The Phenomenon of Man* (New York: Harper, 1959).

massocre of ethnic group

tory of mankind, through (pogroms) and dark ages, through wars and concentration camps. We may believe the universe has a direction, or we may disbelieve it. But we cannot, as reasonable men, ignore the fact that the question of man's importance has been a persistent philosophical enigma. If we cannot prove the importance of persons, and if we cannot reasonably ignore the issue, what are we to do?

Since every culture differs in its estimate of the value of persons, and since this information is transmitted through the Parent, we can find no way of relying on the Parent to come to any agreement on the worth of persons. In many cultures, including our own, killing is condoned by the Parent. The worth of persons, thus, is conditional. In war, killing is acceptable. Capital punishment is legal in many countries, including our own. Infanticide was practiced by many early cultures, with the rationale of preserving the best of the species. The practice of infanticide has been reported even in the twentieth century. For instance, among the Tanala, in Madagascar, there are two groups differing markedly in skin color, although they seem to be much alike in their other physical characteristics and are nearly identical in culture and language. These groups are known by terms roughly translatable as the Red Clan and the Black Clan. Normal members of the Red Clan are a very light brown, and normal members of the Black Clan are a very dark brown. If a dark child of unquestioned clan parentage is born into the Red Clan, it is believed that he will grow up to be either a sorcerer, a thief, a person guilty of incest, or a leper. It is therefore put to death.* This belief about the value of "that kind of person" is passed from generation to generation through the Parent. The cultural Parent of most Western nations does not agree. It does, however, condone other forms of discrimination, which also may end in death.

Neither can we rely on the Child for agreement as to the value

*Ralph Linton, *The Study of Man.*

of persons. The Child, crippled by its own NOT OK, has little positive data about its own value, let alone the value of others. The
Child in any culture, if provoked sufficiently, may break out in
murderous rage, or in murder itself, even mass murder.

Only the emancipated Adult can come to agreement with the
emancipated Adult in others about the value of persons. We can see
how inadequate words such as "conscience" are. We have to ask,
"What is the still, small voice inside us? What is this conscience we
live by? Is it Parent, Adult, or Child?"

Bertrand Russell, never content to let a sleeping dogma lie,
said: "This inner voice, this God-given conscience which made
Bloody Mary burn the Protestants, this is what we reasonable beings are to follow? I think the idea mad and I endeavor to go by reason as far as possible."*

"I Am Important, You Are Important"

The Adult is the only part of us that can choose to make the statement "I Am Important, You Are Important." The Parent and Child
are not free to do so, being committed to that which, on the one
hand, was learned and observed in a particular culture and, on the
other hand, what was felt and understood.

An Adult assertion that Persons Are Important is quite different from the statement made by a woman patient who, while
clenching her fists, said effusively, "I *love* people." This statement
or variations thereof came from her adaptive Child: "Now go kiss
Aunt Ethel, darling!" Four-year-old Darling dutifully does so even
though Aunt Ethel horrifies her. But she does it, and she underlines it, "I *love* Aunt Ethel," and she canonizes it, "I *love* people."
She is still, however, clenching her fists.

We must all examine our own versions of "I love people" to

*B. Russell, *The Autobiography of Bertrand Russell* (Boston: Little, Brown, 1967).

understand how we really feel and where this data is coming from. Most of us claim certain beliefs, but often they are the products of Child acceptance of Parent indoctrination, rather than the conclusions of the Adult on the basis of a purposefully attained body of data.

The Adult's approach to the worth of persons, in contrast, would follow these lines.

I am a person. You are a person. Without you I am not a person, for only through you is language made possible and only through language is thought made possible, and only through thought is humanness made possible. You have made me important. Therefore, I am important and you are important. If I devalue you, I devalue myself. This is the rationale of the position I'M OK— YOU'RE OK. Through this position only are we persons instead of things. The requirement of this position is that we are responsible to and for one another, and this responsibility is the ultimate claim imposed on all men alike. The first inference we can draw is Do Not Kill One Another.

"It Won't Work!"

One afternoon a colleague confronted me in the doctors' parking lot and said gleefully, "If I'm OK and YOU'RE OK, how come you're locking your car?"

The problem of evil also is a reality in the world. In the face of all the evil we see, this fourth position, I'M OK—YOU'RE OK, may seem to be an impossible dream. It may be, however, that our civilization is rapidly arriving at an unprecedented confrontation: we either respect each other's existence or we all perish. And, even in a most detached manner, we would have to say this would be a shame, to bring to an end what has taken so long to build.

Teilhard, who with exquisite wonderment perceives the unfolding of the universe as a refining and converging evolutionary process which is still going on, nevertheless ends his great work *The Phenomenon of Man* on a note of pain as he contemplates the evil in the universe, wondering if perhaps all the suffering and failure, tears and blood, "does not betray a certain excess, inexplicable to our reason, if to the normal effect of evolution is not added the extraordinary effect of some catastrophe or primordial deviation?"

Are we an evolutionary mistake? Or do the remarkable events in the development of man promise still greater preserving events in the future? Teilhard speaks of that moment when the first man reflected, when he knew himself to be, as "a mutation from zero to everything."

Perhaps we are approaching another significant point, where because of the necessity of self-preservation we shall undergo another mutation, we shall be able to leap again, to reflect—with new hope based on the enlightenment of how we are put together—I am important, you are important. I'M OK—YOU'RE OK.

I believe Transactional Analysis may contribute to an answer to the predicament of man. Despite the seeming presumptiveness of this assertion, I take courage from J. Robert Oppenheimer's vision that there be "common discourse, a continuous interplay between the world of scientists and the world of people at large, artists, farmers, lawyers, and political leaders." In 1947 he wrote, " . . . because most scientists, like all men of learning, tend in part to be teachers, they have a responsibility for the communication of the truths they have found." In his view expressed in 1960, men in "high intellectual enterprise must contribute to the common culture, where we talk to each other, not just about the facts of nature . . . but about the nature of the human predicament, about the nature of man, about law, about the good and the bad, about morality, about political virtue, and about politics."*

We have a responsibility to apply our findings in the observa-

*Thomas B. Morgan, "With Oppenheimer," *Look,* Jan. 27, 1966.

tion of the transactions between persons to the broader problem of the preservation of mankind.

The Original Game Is the Original Sin

I believe it is possible from the data at hand to say something new about the problem of evil. Sin, or badness, or evil, or "human nature," whatever we call the flaw in our species, is apparent in every person. We simply cannot argue with the endemic "cussedness" of man. I believe the universal problem is that by nature every small infant, regardless of what culture he is born into, because of his situation (clearly *the* human situation), decides on the position I'M NOT OK—YOU'RE OK, or the other two variations on the theme: I'M OK—YOU'RE NOT OK or I'M NOT OK—YOU'RE NOT OK. This is a tragedy, but it does not become demonstrable evil until the first game is begun, the first ulterior move is made toward another person to ease the burden of the NOT OK. This first retaliatory effort demonstrates his "intrinsic badness"—or original sin—from which he is told he must repent. The harder he fights, the greater his sin, the more skillful become his games, the more ulterior becomes his life, until he does, in fact, feel the great estrangement, or separateness, which Paul Tillich defines as sin.* But what he does (games) is not the primary problem; it is rather what he considers himself to be (the position). Tillich says, "Before sin is an act, it is a state." Before games were played, a position was taken. I am convinced that we must acknowledge that this state—the position I'M NOT OK—YOU'RE OK—is the primary problem in our lives and that it is a result of a decision made early in life under duress, without due process and without an advocate. But as we see the truth of the situation, we can reopen the case and make a new decision.

One of my patients said, "I play a game of 'Internal Court-

*P. Tillich, *The Shaking of the Foundations* (New York: Scribner's, 1950).

room,' with my Parent acting as a judge, jury, and executioner. It's a fixed trial, as my Parent decides in advance that I'm guilty. I never realized that a defendant is entitled to a lawyer. I never attempted to defend my Child. My Parent did not permit me to question its judgment. But my computer finally clicked and made me aware that there is another option open—my Adult can evaluate the situation and intercede for my Child. The Adult is the lawyer."

Through the enlightenment that I'M NOT OK was a wrong decision comes the reprieve after which one can begin to understand it is safe to give up games.

P-A-C and Religion

The Parent-Child nature of many religions is remarkable when one considers that the revolutionary impact of the most revered religious leaders was directly the result of their courage to examine Parent institutions and proceed, with the Adult, in search of truth. It takes only one generation for a good thing to become a bad thing, for an inference about experience to become dogma. Dogma is the enemy of truth and the enemy of persons. Dogma says, "Do not think! Be less than a person." The ideas enshrined in dogma may include good and wise ideas, but dogma is bad in itself because it is accepted as good without examination.

Central to most religious practices is a Child acceptance of authoritarian dogma as an act of faith, with limited, if not absent, involvement of the Adult. Thus, when morality is encased in the structure of religion, it is essentially Parent. It is dated, frequently unexamined, and often contradictory.* I pointed out earlier that since every culture differs in its estimate of the worth of persons, and since this information is transmitted through the Parent, we can find no way of relying on the Parent to come to any agreement as to the

*See James A. Pike, *You and the New Morality* (New York: Harper, 1955).

worth of persons. Thus, Parent morality, rather than advancing the idea of a universal ethic, which has a claim on all men, impedes the formulation of a universal ethic. The position I'M OK—YOU'RE OK is not possible if it hinges on your accepting what I believe.

I am limiting the following observations to the Christian religion, because it is the only religion about which I have enough data to warrant observation. The central message of Christ's ministry was *the concept of grace.* Grace is a "loaded" word, but it is difficult to find a word to replace it. The concept of grace, as interpreted by Paul Tillich, the father of all the "new Christian theologians," is a theological way of saying I'M OK—YOU'RE OK. It is not YOU CAN BE OK, IF, or YOU WILL BE ACCEPTED, IF, but rather YOU ARE ACCEPTED, unconditionally.

He illustrates this by referring to the story of the prostitute who came to Jesus. Tillich said, "Jesus does not forgive the woman, but he declares that she *is* forgiven. Her state of mind, her ecstasy of love, show that something has happened to her." Tillich stated further, "The woman came to Jesus because she was forgiven, not to be forgiven."* She probably would not have come to him had she not known already that he would accept her in love, or grace, or I'M OK—YOU'RE OK.

This concept is incomprehensible to many "religious persons," because it can only be perceived by the Adult, and many religious persons are Parent-dominated. The Parent has too many reservations about the other guy and reads the creed YOU CAN BE OK, IF. The Child, on the other hand, has devised many games to evade the judgment of the Parent. An example of such a game is "Religious Schlemiel," a variation of the game "Schlemiel," described by Berne.† This is a game wherein the Sinner (who is It) goes through the week; foreclosing his tenants, underpaying his employees, belittling his wife, yelling at his children, spreading gossip about his competitors, and then, on Sunday, says a sing-song

*P. Tillich, *The New Being* (New York: Scribner's, 1955).
†Eric Berne, *Games People Play* (New York: Grove Press, 1964).

"I'm sorry" to God, thereupon leaving the church with the assumed assurance that "12 o'clock and all's well!"—which is the
payoff.

Not all "sinners" are such blatant game players. However, because their internal religious dialogue is predominantly Parent-
Child, they are continually caught up in an anxious scorekeeping
of good and bad works, never sure of how they stand. Paul
Tournier states that religious morality "substitutes for the liberating experience of grace [I'M OK—YOU'RE OK] the obsessive fear of
committing a mistake."*

If, like Tillich, we understand our primary problem as a *state*
(an estrangement, a NOT OK position, or sin, singular) and not as
an *act* (acts of sin, games to overcome the position, or sins, plu-
ral), then we see the ineffectiveness of "confession of *sins*," over
and over again, in producing change in a person's life. Tillich
says that for some people, grace is the "willingness of a divine
king and father to forgive over and over again the foolishness and
weakness of his subjects and children. We must reject such a concept of grace; for it is merely a childish destruction of human dignity."† Such a view only adds to the NOT OK. It is the *position*
which we must "confess," or acknowledge, or comprehend. Then
it becomes possible to understand games and to become free to
give them up.

A confession by the Adult is quite different from a confession
by the Child. Whereas the Child says, "I'm sorry . . . I'M NOT OK . . .
please forgive me . . . ain't it awful," the Adult can make a critical assessment of where change is possible and then follow through.
Confession without change is a game. This is true whether in a sanc-
tuary, a pastor's study, or a psychiatrist's office.

The non-Adult transmission of Christian doctrines has been
the greatest enemy of the Christian message of grace. The message
has been distorted throughout history to fit the game patterns of

*P. Tournier, *The Seasons of Life* (Richmond, Va.: John Knox Press, 1961).
†Tillich, *The Shaking of the Foundations.*

every culture to which it has been introduced. The I'M OK—YOU'RE OK message has been twisted again and again to a WE'RE OK—YOU'RE NOT OK position under which sanction Jews have been persecuted, racial bigotry has been established as moral *and* legal, repeated religious wars have been fought, witches have been burned, and heretics have been murdered. The doctrine of grace (I'M OK—YOU'RE OK) is hardly recognizable in such doctrines preached by the Parent-damning and Child-raging Elmer Gantrys and Jonathan Edwardses who saw the glories of heaven in terms of a ringside seat at the right hand of God to watch the spectacle of the damned burning in hell.

This was hard game playing designed to make individuals squirm. Many ministers today "keep their cool" about questions of the sin and repentence of individuals and have turned their attack on the sin of society, in an attempt to make society squirm. This "attack" varies from a mild sociology lecture to an angry assault against social injustice. However, slums and ghettos and put-downs are not going to disappear in society unless slums and games disappear from the hearts of people. A convincing example of this fact is the outcome of the vote on California's Proposition 14 in 1964. This, in effect, was a proposition against fair housing. "Society's" stand was clear: Almost every major organization in the state officially opposed it—almost all religious organizations, boards of education, the major political parties, chambers of commerce, labor unions, the State Bar Association, and the PTA, to name just a few of the representative organizations. Yet the proposition won by a two-to-one vote. What society *should* do is one thing. "What individuals *dare* do is another.

Failure to make a difference in these *relevant* areas has driven many ministers to despair, has caused many to leave the ministry, and has pushed others back into a resigned acceptance of the "conservative view," in which, despite lofty doctrinal pronouncements, the church, in effect, is a repository of Parent dictates designed to keep things as they are—cooperate with the chamber of commerce, pay for the building, baptize, marry, and bury. There is goodness in much of what is done, but, considering the state of the world, these activities are hardly enough. Young clergymen graduating from seminaries today, inspired by Bonhoeffer, Tillich, and Buber, become depressed and disillusioned when they find they have been hired to referee the games of the church, babysit, plan nice social events for the young people, and keep young girls from getting pregnant. The contract is that *we don't really have to change; after all, we are such nice people.* It is highly probable that the historical Jesus would be turned away from many Sunday morning worship services. Jesus was called a winebibber and a glutton, because he enjoyed the communion of ordinary persons. The twentieth-century WASP* Parent says, "You are judged by the friends you keep; don't associate with *their* kind." Jesus said, "Feed my sheep." The Parent says, "That's what we pay our preacher for." Jesus said, "Blessed are the poor and meek." The Child says, "Mine's better than yours." Jesus said the greatest commandment is, "Thou shalt love the Lord thy God with all thy heart and thy neighbor as thyself." The Parent says, "We don't want *them* moving into our neighborhood." The Child is involved, too. The Child is afraid of *them.*

Unfortunately, many of the people whose Adult cannot "buy all that inconsistency and hypocrisy" have, as in the case of the baby and the bath water, dumped Christ's original message along with the muddy waters of "Christianity." It is to the task of restoring the simple message of personal liberation and clearing away the mud of institutional dogma that the "new theologians" have turned their efforts.

If personal liberation is the key to social change, and if the truth makes us free, then the church's principal function is to provide a place where people can come to hear the truth. *The truth is a growing body of data of what we observe to be true.* If Transac-

* White Anglo-Saxon Protestant.

tional Analysis is a part of the truth which helps to liberate people, the churches should make it available. Many ministers who have been trained in Transactional Analysis agree and are conducting courses in Transactional Analysis for members of their churches as well as using it in pastoral counseling.

What Is a Religious Experience?

Is there such a thing as a religious experience, or is such an experience simply a psychological aberration? Does the mind "just get carried away" with a wish, as Freud* suggested, or is there more to it than fantasy?

Trueblood states,

> The fact that a great many people representing a great many civilizations and a great many centuries, and including large numbers of those generally accounted the best and wisest of mankind, have reported direct religious experience is one of the most significant facts about our world. The claim which their reports make is so stupendous and has been made in such a widespread manner that no philosophy can afford to neglect it. Since we cannot hope to build up a responsible conception of the universe unless we take into consideration every order of fact within it, the claim of religious experience with objective reference cannot be lightly dismissed. The very possibility that anything so important *might* be true gives our researches into these matters a necessary mood of seriousness. Since the claim is conceivably valid, there is the heaviest reproach upon those who fail to examine it seriously. When the claim is cavalierly rejected, without a careful examination; this is presumably because of some dogmatic position.[†]

*See Ernest Jones, *The Life and Work of Sigmund Freud,* Volume 3 (New York: Basic Books, 1957), pp. 349–360.
†Elton Trueblood, *Philosophy of Religion* (New York: Harper, 1957).

The capacity to reflect on religious experience is significant in itself. Where does our ideation of God, or "the more," or transcendence, come from? Does the God-idea simply grow out of fear of the unknown? Was religious experience reported in the beginning in order to manipulate others by claiming other-worldly powers? Has the God-idea simply evolved, survived because it is somehow related to the survival of the fittest?

Teilhard in *The Phenomenon of Man* takes issue with this view of evolution:

> We are definitely forced to abandon the idea of explaining every case simply as the survival of the fittest, or as a mechanical adaptation to environment and use. The more often I come across this problem and the longer I pore over it, the more firmly is it impressed upon me that in fact we are confronted with an effect not of external forces but of psychology. According to current thought, an animal develops its carnivorous instincts because its molars become cutting and its claws sharp. Should we not turn the proposition around? In other words, if the tiger elongates its fangs and sharpens its claws is it not rather because, following its line of descent, it receives, develops, and hands on the "soul of a carnivore"?

It would appear that something in the development of man has changed, which first appears as the ideation of transcendence, and then as transcendence itself.

Teilhard says further in the same book:

> The law is formal. We referred to it before, when we spoke of the birth of life. No size in the world can go on increasing without sooner or later reaching a critical point involving some change of state.

The first remarkable change of state in the development of man occurred when he crossed the threshold of reflection, what

Teilhard calls a critical transformation, a "mutation from zero to everything." With the power of reflection the cell has become "someone." He said this threshold had to be crossed at a single stride and that it was "a transexperimental interval about which scientifically we can say nothing, but beyond which we find ourselves transported onto an entirely new biological plane."

In view of the "impossible, unprecedented" development of thinking man, is it not reasonable to conclude that there may have developed an "impossible, unprecedented" *transcendent man?*

Transcendence means an experience of that which is more than myself, a reality outside of myself, that which has been called The Other, The All, or God. It is not a "floating upward," as in pre-Copernican paintings; in fact, it is better expressed in the image of depth. This is the way Tillich comprehends it in *The New Being:*

> The name of this infinite and inexhaustible depth and ground of all being is *God.* That depth is what the word *God* means. And if that word has not much meaning for you, translate it, and speak of the depths of your life, of the source of your being, of your ultimate concern, of what you take seriously without any reservation. Perhaps, in order to do so, you must forget everything traditional that you have learned about God, perhaps even the word itself. For if you know that God means depth, you ~~DEPTH~~ know much about him. You cannot then call yourself an atheist or unbeliever. For you cannot think or say: Life has no depth! Life is shallow. Being itself is surface only. If you could say this in complete seriousness, you would be an atheist; but otherwise you are not.

What happens, then, in a religious experience? It is my opinion that religious experience may be a unique combination of Child (a feeling of intimacy) and Adult (a reflection on ultimacy) with the total exclusion of the Parent. I believe the total exclusion of the Parent is what happens in *kenosis,* or self-emptying. This self-emptying is a common characteristic of all mystical experiences, according to Bishop James Pike:

> As we have seen there is a generic character to the mystical experience of, say, a Christian and a Zen Buddhist, and in the experiential patterns of persons of both traditions, can be observed common factors. This is illustrated by the fact that present-day Zen Buddhist philosophers use the same Greek word as is used by both Paul and Western theologians to describe a process which experience—in East and in West—has been found to be a principal route to the consummation of personal fulfillment. The word is *kenosis,* that is, self-emptying.*

I believe that what is emptied is the Parent. How can one experience joy, or ecstasy, in the presence of those recordings in the Parent which produced the NOT OK originally? How can I *feel* acceptance in the presence of the earliest *felt* rejection? It is true that Mother was a participant in intimacy in the beginning, but it was an intimacy which did not last, was conditional, and was "never enough." I believe the Adult's function in the religious experience is to block out the Parent in order that the Natural Child may reawaken to its own worth and beauty as a part of God's creation.

The little person sees the Parent as OK, or, in a religions vein, righteous. Tillich says, "The righteousness of the righteous ones is hard and self-assured." (This is the way the little person sees his parents, even if in fact the parents, according to other standards, are not righteous.) Tillich asks, "Why do children turn away from their righteous parents, and husbands from their righteous wives, and vice versa? Why do Christians turn away from their righteous pastors? Why do people turn away from righteous neighborhoods? Why do many turn away from righteous Christianity and from the Jesus it paints and the God it proclaims?

*James A. Pike, *If This Be Heresy* (New York: Harper & Row, 1967).

Why do they turn to those who are not considered to be the righteous ones? Often, certainly, it is because they want to escape judgment."* (The religious experience is the escape from judgment, acceptance without condition.) The "faith of our fathers" is not the same as *my* faith, although in exercising *my* faith, I may discover the same experience they did, with the same object they did.

There is one kind of religious experience which may be qualitatively different from the Parent-excluding experience we have just described. This is the feeling of great relief which comes from a total adaptation to the Parent. "I will give up my wicked ways and be exactly what you [Parent] want me to be." An example is a "converted" woman whose first act to confirm her salvation is to wipe off her lipstick. Salvation is not experienced as an independent encounter with a gracious God but as gaining the approval of the pious ones who make the rules. The "will of God" is the will of the congregational Parent. Freud believes religious ecstasy is of this sort: The Child feels omnipotence by selling out to the omnipotent Parent. The position is I AM OK AS LONG AS. The conciliation produces such a glorious feeling that there is a hunger for it to happen again. This results in "backsliding," which paves the way for another "conversion" experience. The Adult is not involved in this experience. The religious experience of children may be of this sort. We cannot be judgmental about religious experiences of others for there is no certain, objective way to know what really happens to them. We cannot say that one person's experience is genuine and another's is not. A subjective appraisal, however, leads me to believe that there is a difference in a religious experience based upon Parent approval and a religious experience based on acceptance without condition.

If it is true that we empty ourselves of the Parent in the religious experience first described, this leaves the Child and Adult. Whether

*Tillich, *The New Being.*

God is experienced by the Child or by the Adult is a fascinating question. It has been said that the God of the philosophers is not the same God as the God of Abraham, Isaac, and Jacob. The God of the philosophers is a "thought" construction, an Adult search for meaning, a reflection about the possibility of God. Abraham, Isaac, and Jacob "walked with God and talked with God." They *experienced* transcendence. They felt it. Their Child was involved.

Theology is Adult. Religious experience involves the Child, also. It may be that religious experience is totally Child. After all, the Abraham who followed God out of the Land of Ur had not read the Torah, and Paul was converted without the benefit of the New Testament. They reported an experience, and their lives changed because of that experience.

"That which we have seen and heard declare we unto you," wrote John. Perhaps the spontaneity and vigor of the early church was due to the fact that there was no formal Christian theology. Early Christian literature was essentially a report of what happened and what had been said. "Once I was blind and now I see" is a statement of an experience and not an interesting theological idea. The early Christians met to talk about an exciting encounter, about having met a man, named Jesus, who walked with them, who laughed with them, who cried with them, and whose openness and compassion for people was a central historical example of I'M OK—YOU'RE OK.

H. G. Wells said, "I am an historian. I am not a believer. But I must confess, as an historian, this penniless preacher from Galilee is irresistibly the center of history."

The early Christians trusted him and believed him, and they changed. They talked to each other about what happened. There was little of the ritualistic, nonexperiential activity so characteristic of churches today. Dr. Harvey Cox of Harvard Divinity School said in an interview with *Colloquy,* a monthly magazine published by the United Church of Christ:

The earliest gatherings of the followers of Jesus ... lacked the cultic solemnity of most contemporary worship. These Christians gathered for what they called the breaking of bread—that is, the sharing of a common meal.

They had bread and wine, recalled the words of Jesus, read letters from the Apostles and other groups of Christians, exchanged ideas, sang, and prayed. Their worship services were rather uproarious affairs ... more like the victory celebrations of a football team than what we usually call worship today.*

Theirs was a new, revolutionary style of life based on mutual belief and acceptance of each other. If Christianity were simply an intellectual idea, it probably would not have survived, considering its fragile beginnings. It survived because its advent was an historical event, as was Abraham's leaving the Land of Ur, as was Moses' exodus from Egypt, as was Paul's conversion on the road to Damascus. We may not understand religious experience, we may differ in its explanation, but we cannot, if we are honest, deny the reports of such experiences by reputable men through the centuries.

People in Perspective

One of the most helpful ways to strengthen the Adult for the task of examining Parent data (which data can be extremely overpowering to the Child, particularly on the subject of faith) is to stand back for greater perspective, the broader view.

Faith, Trueblood says, is not a blind leap into nothing but a thoughtful walk in the light we have. Part of that light is the recognition that the world which "God so loved" is considerably larger than our own personal comprehension of it. If nothing else, this

---

* "Worship: Clack or Celebration—An Interview with Harvey Cox," *Colloquy,* Vol. 1, No. 2 (February 1968).

recognition should make us modest and rule out our claims to exclusive truth.

I am reminded of one politician's statement, "When white and black and brown and every other color decide they're going to live together as Christians, then and only then are we going to see an end to these troubles." That statement may mean something to him; but what does it mean to the one and a half billion persons in the world today who don't know who Christ was and never have heard his name?

This brings us to a way in which we can look at the world's people in perspective. In a sermon I heard some time ago the following statistics were presented:

If the three billion people of the world could be represented in a community of one hundred:

Six would be United States citizens; ninety-four would be citizens of other countries.

Six would own one-half of the money in the world; ninety-four would share the other half; of the ninety-four, twenty would own virtually all of the remaining half.

Six would have 15 times more material possessions than the other ninety-four put together.

Six have 72 percent more than the average daily food requirement; two-thirds of the ninety-four would have below-minimum food standards, and many of them would be on a starvation diet.

The life span of six would be seventy years. The life span of ninety-four would be thirty-nine years.

Of the ninety-four, thirty-three would come from countries where the Christian faith is taught. Of the thirty-three, twenty-four would be Catholic and nine would be Protestant.

Less than one-half of the ninety-four would have heard the name of Christ, but the majority of the ninety-four would know of Lenin.

Among the ninety-four there would be three communist documents which outsell the Bible.

By the year 2000 one out of every two persons will be Chinese.*

We are deluded if we continue to make sweeping statements about God and about man without continually keeping before us the facts of life: the long history of the development of man, and the present-day diversity of human thought. This may be frightening data to some people. "Hopeless!" they may cry. I rather like Teilhard's view. When asked once what made him happy, he said: "I'm happy because the world is round." The borders, corners, or angles are not physical, but psychological. If we remove the psychological fences erected to protect the NOT OK Child existent in every person, there are no barriers to prevent our living together in peace.

What Is Reality Therapy?

Early in this chapter I stated that reality is our most important treatment tool. I have proceeded to discuss a number of realities. In concluding this chapter I wish to compare briefly Transactional Analysis with Reality Therapy, developed by Dr. William Glasser.[†] Glasser holds that man's basic problem is moral in that *being responsible* is the requirement for mental health.

I believe both approaches—Transactional Analysis and Reality Therapy—can be thought of as products of a new breakthrough in psychiatry born of the dissatisfaction with the ineffectiveness and unreality of those types of psychiatry and clinical psychology, which, in effect, dismiss morality from the focus of treatment. Both Transactional Analysis and Reality Therapy hold that people are responsible for their behavior. There is an essential difference, however. I disagree with Glasser in his general denial of the significance of the past in understanding behavior in the present. I do

*Michael D. Anderson, sermon, Fremont Presbyterian Church, Sacramento, Dec. 27, 1964.
[†]W. Glasser, *Reality Therapy* (New York: Harper & Row, 1965).

not believe in the game of "Archaeology," or digging in the past, but neither do I believe we can totally ignore the past. To me the man who ignores his past is like the one who stands in the rain, arguing about its wetness while becoming drenched. Telling a patient he must be responsible is nowhere near the same thing as his becoming responsible. Transactional Analysis is also a "reality therapy," but it provides answers that I do not believe Glasser has provided. What is wrong with people, for instance, who cannot perceive reality or whose perception is distorted (contaminated)? What is the answer to those who "know what they must do but continually fail to do it"?

Glasser states, "We do not concern ourselves with unconscious mental processes . . . we do not get involved with the patient's history because we can neither change what happened to him or accept the fact that he is limited by his past."

It is true we cannot change the past. Yet the past invariably insinuates itself into our present life through the Parent and the Child, and unless we understand why this happens, and admit that it does, we do not have an emancipated Adult by which we can become the responsible persons Glasser admonishes us to be. We have to understand our P-A-C before we can turn off the past. When a therapist tells us we *must,* this is Parent. If we choose to do so ourselves because we understand how we are put together, this is Adult. The "staying power" of our decision is totally dependent upon whether the decision is Parent or Adult.

Another reservation I have about Reality Therapy is that it does not have a special language with which to report "what happened." Glasser states, "The ability of the therapist to get involved is the major skill of doing Reality Therapy but it is most difficult to describe. How does one put into the words the building of a strong emotional relationship quickly between two relative strangers?"

In Transactional Analysis we have these words. The patient in the beginning comes on Child and views the therapist as Parent. In the initial hour, Parent, Adult, and Child are defined, and these words are then used to define the contract, or mutual expectations from treatment. The therapist is there to teach and the patient is there to learn. The contract is Adult-Adult. If the patient is asked, "What happened?" he can tell what happened. He has learned to identify his own Parent, Adult, and Child. He has learned to analyze his transactions. He has acquired a *tool* to free up and strengthen his Adult, and only this Adult can be responsible.

I agree wholeheartedly with Glasser's central focus of *responsibility*, just as I agree with the ideal of the Ten Commandments and the Golden Rule. The reality that concerns me, however, is why these admonitions do not routinely produce responsible persons. To simply restate them in new ways is not going to do the job.

We cannot produce responsible persons until we help them uncover the I'M NOT OK—YOU'RE OK position which underlies the complicated and destructive games they play. Once we understand positions and games, freedom of response begins to emerge as a real possibility. As long as people are bound by the past, they are not free to respond to the needs and aspirations of others in the present; and "to say that we are free," says Will Durant, "is merely to mean that we know what we are doing."*

---

*W. Durant, *The Story of Philosophy* (New York: Simon and Schuster, 1963), p. 339.

13
Social Implications
of P-A-C

History is populated by tyrants who have done the
inconceivable. And the button exists.

—*In Search of Man*, **Documentary by ABC-TV
and Wolper Productions**

Does our understanding of why individuals act as they do throw any light on why groups of people, such as nations, act as they do? It is important that we ask this question, because if it is not asked and answered soon, there may be little point in being concerned about individuals.

"Do you really think a human being is a rational being?" Senator William Fulbright asked at a Senate Foreign Relations Committee hearing. "In Vietnam," he continued, "in order to give an election to a people that never had an election we are willing to kill thousands of them. This seems to me irrational."

Since collective as well as personal modes of behavior are transmitted from one generation to another through the Parent, it is important for a nation to be as scrupulously critical of its existing institutions and procedures as it is for an individual. The United

States affords great freedom for this kind of critical examination, and yet there is a question of how effectively and truthfully we exercise this freedom. We defend our national, or collective, Parent sometimes rather blindly and seem to forget that other nations do the same thing. We call our defense "patriotism," and their defense "enslavement." To some extent all nations live behind a curtain. Perhaps it is the same curtain.

The California Superintendent of Schools, Max Rafferty, defines good citizenship in this way:

> The good citizen stands in relation to his country as the good son to his mother.
>
> He obeys her because she is his elder, because she conjoins within herself the vision of many, and because he owes to her his begetting and his nurturing.
>
> He honors her above all others, placing her in a special niche within his secret heart, in front of which the candles of respect and admiration are forever kept alight.
>
> He defends her against all enemies, and counts his life well lost in her behalf.
>
> Above all else, he loves her deeply and without display, knowing that although he shares that privilege with others, the nature of his own affection is unique and personal, rising from the deepest well-springs of his being, and returned in kind.
>
> This is the good citizen. While his kind prevails, so also flourishes the Great Republic.*

The only thoughtful response to such a pronouncement is, "That depends." Whether we obey, honor, and defend our mother, our Parent, or our national Parent, depends on what this Parent really is. It may be that because we feel we must believe in an idea, we cannot see what the idea is.

This is precisely the same kind of devotion which makes the people of India allow the rats to eat 20 percent of their inadequate

*Max Rafferty, *California Education,* Vol. II, No. 8 (April 1965).

food supply, or makes an Indian woman bear ten children to starve in the streets because her Parent will not let a male doctor install an intrauterine contraceptive device, now being mass-produced in India. Her Parent does not object to the device, only to the male doctor. There are not enough female doctors to perform this procedure on a large scale. Throughout the world we see evidences of "blindness," and yet we fail to see that it is a blindness common to all men. It is the same blindness as that of the little boy in Chapter 2, who must believe "cops are bad" in the face of contrary evidence supplied by his own eyes and ears. It is the original fear and dependency in the little child which makes it imperative to accept the parents' dictates for the preservation of his life. We can look at his predicament with sympathy. Perhaps if we concentrate not on the Parent of our "world enemies" but rather on their Child, with the hope of re-establishing Adult-Adult communication, we can begin a sympathetic rather than a frantic appraisal of what can be done to work in the direction of a better world.

Our own fears condition us, for instance, to see only the ominous Parent of Red China, threatening, foreboding, angry, and strong. A different point of view is expressed by Eric Sevareid in his estimate of the position taken by Senator William Fulbright regarding Red China:

> Fulbright as a student of history and its unpredictability would find such fears childish. He is more inclined to interpret China's thunderous propaganda challenges as Secretary General U Thant of the United Nations does—as the natural behavior of a regime that is overwhelmed with difficulties at home and feels increasingly "encircled" by the power of Russia and the United States. Fulbright's mental processes are such that he would try to imagine the reaction of his own country if a Chinese army were fighting, say, in lower Mexico, and their planes were dropping bombs within forty miles of the Rio Grande.
>
> He tries to turn an international problem around, not only to understand an adversary's basic interests, but to try to imagine

how the adversary feels in his heart. He thinks the world is too dangerous to do otherwise.*

To Fulbright's question of whether man is rational, Dr. Jerome Frank, Professor of Psychiatry at Johns Hopkins University, who was present at the Senate Foreign Relations Committee hearing, replied: "We are rational only by fits and starts. I think we operate under a great deal of fear and emotional tension, which interferes with clear thinking. We have a right to be afraid of nuclear weapons."

The little child also has a right to be afraid of a beating from a brutal father. The more relevant consideration, however, is not whether or not he has a right to be afraid, but what he can do about it. When fear dominates his life, there is no possibility for the kind of precision data processing which can make possible a position of cure (individual or worldwide), I'M OK—YOU'RE OK.

This was expressed on another occasion by Senator Fulbright in a 1964 speech (interpolations in brackets are the author's):

> There is an inevitable divergence, attributable to the imperfections of the human mind [the Contaminated Adult], between the world as it is [viewed by the Emancipated Adult] and the world as men perceive it [viewed by the Parent or Child or the Contaminated Adult]. As long as our perceptions are reasonably close to objective reality [uncontaminated], it is possible for us to act upon our problems in a rational and appropriate [Adult] manner. But when our perceptions fail to keep pace with events [are archaic], when we refuse to believe something because it displeases [Parent] us or frightens [Child] us, or because it is simply startlingly unfamiliar, then the gap between fact and perception becomes a chasm and action becomes irrelevant and irrational. . . .†

*An interview with Senator William Fulbright, *The Congressional Record,* April 20, 1966.
†W. Fulbright, "Foreign Policy—Old Myths and New Realities," *The Congressional Record,* March 25, 1964.

How Irrational Are We Capable of Being?

The horror most people felt over the disclosures of what happened in Nazi Germany during World War II was accompanied, not infrequently, by the self-righteous assumption that "this could never happen here," that *we* could never be capable of allowing such incredible atrocities.

✓

Could we never? What happened in Nazi Germany? Are all people capable of being irrational? How irrational? Who draws limits?

One of the most chilling articles I have read recently appeared as a book review by psychiatrist Ralph Crawshaw of *The Corrupted Land: The Social Morality of Modern America* by Fred J. Cook, published by Macmillan. Crawshaw wrote:

> Essentially, Cook is saying in *The Corrupted Land* that American citizens have abandoned their personal morality for a collective, institutionalized morality. They have abandoned thoughtful conviction for compromised sentimentality and popularity, that is, responsibility for obedience. This is strong medicine to take. We can always hide behind the fact that he has no statistical evidence, that it is his impression, so it really does not matter much anyway. Or does it?*

I quote at length from the review, wherein Crawshaw reports a research project conducted by Stanley Milgram at Yale University that provides evidence in answer to this question:

> Stanley Milgram's research at Yale begins to put a scientific foundation beneath Cook's deductions. Milgram carried out a series of psychological experiments in obedience. He took a sample cross-section of the adult (aged twenty to fifty) male

*Ralph Crawshaw, "But Everybody Cheats," *Medical Opinion and Review,* Vol. 3, No. 1 (January 1967).

population of Bridgeport, Connecticut, from blue-collar workers through professionals. The experiment was to determine
how much punishment one person would inflict on another
when so ordered, but it was deceptively presented as a scientific
study by "Bridgeport Research Associates" on learning techniques. The subjects were selected at random, paid four and a
half dollars for their time.

A learner was strapped into an "electric chair" after the subject (a teacher) had experienced a sample shock of 45 volts to
convince him of the authenticity of the instrument. Then from
an adjoining room the subject (teacher) delivered a shock of increasing intensity for each wrong answer. Actually the learner
was not shocked but his responses were a standardized tape
recording of murmurs, gasps, pleas, and shrieks matched to the
ascending voltage. The answers were prearranged by the experimenter, so the only limiting factor preventing the teacher from
delivering the maximum intensity of the shock machine was his
compassion for a human being in pain. The conflict for the
teacher (Mr. Everyman) was obedience to the authoritative system versus his belief that he should not hurt another person.

What percentage of the so-called teachers went the full route
and gave the maximum shock? But before you guess, listen to
the voice of Mr. Everyman at the switch:

150 volts delivered: "You want me to keep going?" (COM-
MAND)

165 volts delivered: "That guy is hollering in there. There's a
lot of them here. He's liable to have a heart condition. You want
me to go on?" (COMMAND)

180 volts delivered: "He can't stand it. I'm not going to kill
that man in there! You hear him hollering? He's hollering. He
can't stand it. What if something happened to him? You know
what I mean? I mean, I refuse to take the responsibility." (THE
EXPERIMENTER ACCEPTS THE RESPONSIBILITY) "All right."

195 volts, 210, 225

240, and so on.

The subject (teacher) unfailingly obeyed the experimenter.
What percentage of the almost one thousand teachers went the

whole route? Estimate before you let your eyes stray ahead. A group of forty psychiatrists who studied the project predicted one-tenth of one percent. In the actual experiment, sixty-two per cent obeyed the experimenter's commands fully. What did you guess?

Milgram concluded: "With numbing regularity good people were seen to knuckle under the demands of authority and perform actions that were callous and severe. Men who in everyday life are responsible and decent were seduced by the trappings of authority, by the control of their perceptions, and by the uncritical acceptance of the experimenter's definition of the situation into performing harsh acts. The results, as seen and felt in the laboratory, are to this author disturbing. *They raise the possibility that human nature, or more specifically, the kind of character produced in American democratic society, cannot be counted on to insulate its citizens from brutality and inhumane treatment when at the direction of a malevolent authority."* [Italics mine.]

The implications of this experiment are indeed frightening if we view the results as having only to do with something irredeemable in human nature. However, with Transactional Analysis we can talk about the experiment in a different way. We can say that 62 percent of the subjects did not have a freed-up Adult with which to examine the authority in the Parent of the experimenters. Undoubtedly one unexamined assumption was: Whatever experiments are necessary for research are good. This is perhaps the same assumption that helped "reputable" scientists participate in the laboratory atrocities in Nazi Germany.

As little children most of us were taught "proper respect" for authority. This authority resided in the policeman, the bus driver, the minister, the teacher, the postman, the school principal, and also in the faraway personages of the governor, the congressman, the general, and the President. The response of many persons to the appearance of these authority figures is *automatic.* For in-

*S. Milgram, *Human Relations,* Vol. 18, No. 1 (1965).

stance, if you are driving fast and suddenly spot a highway patrol car you do not consciously reason that you had better slow down; your foot automatically lifts from the accelerator. The old "You'd Better Watch Out" recording comes on full volume, and the Child automatically responds, as it always has. On reflection, the Adult recognizes that speed laws are necessary. Therefore, the automatic response is good in this situation.

Not all automatic responses to authority are good. There may be great risk in compliance if the Adult fails to process new data in a changing world. Therefore, in spite of our trepidation, we may view with hope the present climate of dissent and inquiry in our country. The demonstrations and hard questions of the young indicate there is health and strength in their unwillingness to buckle under blindly to authority or to accept without question laws which they consider inimical to justice and survival. Laws are not ultimate truth. There have been bad laws along with the good, and many bad ones have been changed as a result of protest of the kind we see today. If we do not take into account nonviolent protest, we may expect increasing evidence of a take-over by the Child in rioting and violence. If we do not respond to reason, our responses will more and more be dominated by fear. At the same time we must take into account the requirements of the democratic process, which cannot function without laws. As Churchill said, "Democracy is the worst form of government one can imagine until one tries to imagine one that is better." But democracy can only function with an intelligent electorate, and an intelligent electorate is an Adult electorate. A government of the Parent, for the Parent, and by the Parent will perish from the earth.

Is This Younger Generation Different?

Many parents are sorely troubled over the independent ways of today's youth. The thought of letting up on parental pressures is not a welcome one. If anything, some say, we must put the pressure on harder. It is impossible for many parents to believe that anything constructive or practical can reside in the head of a young collegian who wears long hair and protest buttons, and who smokes marijuana, even though those same parents may not be able to build any impressive case for their own short hair, the initiation rites of their fraternal organization, or their cocktail party rituals. "But these spoiled kids are destroying everything we have worked so hard to build," said one irate father regarding the free-speech movements at the University of California at Berkeley. There is truth in this statement; young people can be and some are destructive. They have not paid taxes, they have not helped to build the institutions they attack. On the other hand, they cannot vote, yet they are asked to pay more than taxes. They are required to give their lives in wars which many of them do not support.

A P-A-C examination of today's college student provides a new understanding of his character, which, I believe, helps us lift this subject out of the classical contest (the older versus the younger generation), with its hand-wringing and nonproductive "Ain't It Awful."

In 1965 one of the world's great institutions of learning, the University of California at Berkeley, was rocked by a series of noisy transactions, which were broadcast throughout the world. Much in evidence was the rebellious Child of many of the students in such slogans as "Don't trust anyone over thirty." The Parent was also in evidence, as in the righteous indignation of the Chairman of the Board of Regents over the flagrant use of *the* impolite four-letter word. Also in evidence was the impressive Adult of the then President of the University, Clark Kerr, who was fired in January of 1967. (Decisions made by the Adult do not guarantee acclaim, popularity, or safety, particularly among those who are too threatened by reality to take a second look at it.)

What was really happening on the Berkeley campus? What were the real meanings of the four-letter word? Why in an institution known to be one of the most liberal in the nation were students demanding unlimited freedom in a noisy and blatant protest against all university authority? In a comprehensive analysis of the Berkeley uprising *Fortune's* Max Ways commented:

> Never did an educational institution less deserve the name of tyrant than the University of California. Students can—and at Berkeley most do—live off campus without any university supervision of their conduct. The range of academic choice is huge and only lightly trammeled by curriculum requirements. Indeed, many of the Berkeley student complaints, verbalized as demands for more freedom, derive in fact from the consequences of what educators in less advanced universities would regard as excessive freedom.*

He further observed that "insufficient previous exposure to the institutional type of authority, which works through impersonal rules, makes the university—and the society—appear to many students as a tyrannical Establishment.

The idea of *previous exposure* is an important one. Let us examine the first five years of most undergraduate students, many of whom, if not active in the student rebellion, were sympathetic to it. The age range of undergraduate college students is eighteen to twenty-two. Many of the student protesters were born from 1943 to 1946, and their most formative years were spent either partly during wartime or in the years immediately after the war. These years were characterized by unstable family constellations, movement from place to place, absent if not dead fathers, anxious, weary, troubled mothers, and general social patterns which magnified the unrest in the home. Many young fathers, returning from battle, entered colleges and universities under the G.I. Bill and reflected soberly on the state of a world which had demanded so much of

---

*M. Ways, "On the Campus: A Troubled Reflection of the U.S.," *Fortune,* September-October 1965.

them. Their Purple Hearts and wounded spirits supported their verbal expressions of the hatred of war and devastation. They did not capitulate easily to dead institutions and old clichés about how the world should be. Their little children, now collegians, did not see life as a haven of domestic tranquility or a world safe for democracy. They saw, at an early age, the pictures of the concentration camps and registered the serious questions these pictures raised about the goodness of man. This data was recorded in the Parent.

On the other hand many of these children were the recipients of the symbols of affluence which their parents heaped upon them. They were scrubbed, cavity-free, vitamin-filled, orthodontally wired, and insured for a higher education. Yet all of these ministrations did not erase the early recordings, which we now hear playing in the "unreasonable" activities of protesting students. We must be careful to point out that we cannot generalize about all students, or all protesting students. There are certainly exceptions. Some of the protesters were older than others. Some came from homes which remained stable throughout the war years. Nevertheless, this type of analysis is valuable. It is through this type of inquiry that we can get beyond the "Ain't It Awful" about the younger generation.

The early exposure to difficulties and inconsistencies does not mean that these young people can escape the responsibility for their behavior. However, an understanding of what was recorded in the Parent and in the Child of these students helps to make their attitudes understandable. We recognize that archaic data not only comes from the rebellious, anxious Child but also from the Parent, the content of which also contains many imprints of anxiety and rebellion, mistrust and weariness of a world which seems unable to exist very long without war. A significant number of students, many of whom had never lived with an authority they could trust or one which they could not manipulate, were now ready to protest against all authority, including the authority of the University. They had been conditioned to receive a great deal in terms of material comforts but not enough of those evidences which support the assertion that persons are important and that life has purpose. Their Parent is fragmented, their Child is depressed, and their Adult asks urgently, "Isn't there something more?"

One of the several criticisms expressed throughout the controversy at the University was "the University had grown too big"; a similar observation could be made about the population of the world. UCLA Chancellor Franklin Murphy, who is a physician, replied to this question with a striking biological metaphor, which is also a significant observation about a world which has "grown too big":

> No, it's not too big. But it has had to grow very fast in recent years. The preoccupation has been with the anatomy of the beast rather than with its physiology. If the body gets ahead of the nervous system, the animal gets incoordinate—the animal staggers sometimes. With the university we now have to create a nervous system to match the animal. It takes a sophisticated nervous system to deal with complexity, to carry the messages between the differentiated organs. The university needs more and better decentralization, and it needs more and better coordination.

The function of the "nervous system" of a university is the same as that of the nervous system in the human body—*communication*. It is also the function of the nervous system which ties the world together, and it is the concentration on communication—what facilitates it and what stops it—which will produce something new under the sun rather than the ancient recourse to violence, which is the same whether we call it war, police action, or armed intervention.

Problem solvers, whether international or hometown, talk incessantly about the "need for a dialogue" without ever considering the need to define terms. In Transactional Analysis we have developed a system unique in (1) its definition of terms and (2) its reduction of behavior to a basic unit for observation. Dialogue, if it is to get us anywhere, must be based on agreement of what to examine and an agreement on the words to describe what we observe. Otherwise we simply stumble over words. A person who had known Sirhan Sirhan reported: "He was a fanatic about his country, about political things—but, no, he was not unstable." Words like "fanatic" and "unstable" are useless in analyzing or predicting behavior. Many of our dialogues are useless for the same reason. Much is said, but nothing is understood.

Analyzing International Transactions

If Transactional Analysis makes it possible for two persons to understand what is going on between them, can the same language be used to understand what is going on between nations? As with individuals, the transactions between nations can be complementary only if the vectors on the transactional diagram are parallel. Adult-to-Adult transactions are the only complementary transactions which will work in the world today in view of the self-emergence and self-determination of even the smallest nations. What once was a workable Parent-Child relationship between large and smaller countries is no longer complementary. The smaller countries are growing up. They do not want to be the Child any more. To their sometimes bitter criticisms we respond: How can they feel this way after all we have done for them?

One of the most hopeful institutions for the analysis of international transactions is the United Nations. It has survived many crossed transactions. When the premier of a major nation pounds his shoe on the table, communication stops. When we are told that "they will bury us," it hooks our Child. But we do not have to respond with our Child. Nor do we have to respond with our sword-rattling Parent *And therein lies the possibility of change.*

One has to tell a little child over and over again "I love you," but one "I hate you" is all that is needed for a lifelong negation of any further loving parental advances. If the little person could understand where the "I hate you" came from—how the Child in his parent had been provoked to such an unreasoned and destructive display to the little child he really cherished—then that little child would not have had to hang on to this pronouncement as ultimate truth.

So it is with the "we will bury you" statement of Nikita Khrushchev. Although it was a rather nasty statement and promoted nothing constructive for his country or anyone else's, it may take some of the sting out of it to remember he is only a human being, with a Parent, Adult, and Child, the content of which is different from the Parent, Adult, and Child of anyone else, particularly that of American statesmen.

Proof that he was not a political superbeing is the fact that he himself now is buried, politically. It does not take much historical research to uncover equally blundering statements—and actions—by leaders of other countries, including our own. We must learn to respond to statements and actions of others not with our collective frightened, ready-to-fight Child, but with our Adult, which can seek out the truth, see the fear in the Child of others for what it is, and comprehend the pain they feel from a cultural Parent which dictates absolutes no longer working in the interest of the survival of mankind. We must be able also to stand off at arm's length and view our American cultural Parent. There is much greatness in that Parent, but there is also much wickedness, as in the cancerous evil of slavery, which confronts us now in the murderous faces of racial bigots, both black and white. Elton Trueblood writes:

> In our present predicament, we seem like chess players who have been maneuvered into positions in which all of the available moves are really damaging ones. We begin to see dimly that what is occurring, in part, is a terrible working of moral law, but this is hard for us to believe or to understand. The resentment of Asians seems unreasonable and unjustified, and it is unjustified, if we are concerned only with contemporary events, but

what we are now reaping is a delayed harvest. Every white man who violated, in years past, the principle of the innate dignity of every human being by saying "boy" to a Chinese man, was helping to build up the fierce hatred which finally has burst upon us with such apparent unreason, at the very time that we are trying to maintain a lofty principle at great sacrifice to ourselves.*

Another way of stating this moral law is that if one humiliates the Child in another person long enough he will turn into a monster. It should not surprise us that endless years of humiliation have produced many "monsters" in America who terrify us.

A Negro woman after the Watts riot in Los Angeles responded to all the profuse explanations of why Watts happened (i.e., police action, unemployment, poverty, etc.) by saying: "If they have to ask why, then they'll never know."

I think we would all *know* why if our terrified Child and our self-righteous Parent were not crowding out the Adult.

What to do is another thing. I think we must start by adopting a common language which applies to human behavior, and I feel we have this language in Transactional Analysis. Psychology is heralded as the great "science" of the modern day, yet it has had very little to say which makes sense regarding our present social struggles. To the questioning of the Senate Foreign Relations Committee, another expert in the field of behavior and communications responded, "I'm out of my depth. I feel I'm very liable to make a damn fool of myself." This virtuous display of modesty does not override the fact that those who claim some insight into human behavior ought to have something to say about our relationships with people of other countries.

It is my hope that Senator Fulbright and all of our public officials may receive something more helpful from the psychiatric community. I feel that an understanding of P-A-C and the possibility of the emancipation of the Adult in government leaders and vot-

*E. Trueblood, *The Life We Prize* (New York: Harper, 1951).

ers alike would make one of the greatest contributions to an understanding of the social and world problems that confront us.

By understanding the hold the Parent has on us (our own personal Parent reinforced by a cultural Parent), by understanding the terror in our own Child in the face of rioting and war, in the people of India victimized by starvation and superstition, in the people of Russia in the memory of chains and insurrection, in the people of Israel, reflecting on the most recent murder of six million Jews, in the people of Vietnam, north or south, in the fear of napalm and bayonets, in the people of Japan remembering the A-bomb—if we can begin to see this Child as a little human being in a world full of terror, wanting only the release from pain, then perhaps our international conversations would begin to sound a little different Longfellow suggested that "if we could read the secret history of our enemies we should find in each man's life sorrow and suffering enough to disarm all hostility."

We cannot be sympathetic with the NOT OK Child in our enemies because we are terrified of the games they play to deny the position. And they cannot be sympathetic to us for the same reason. We share the dilemma of mistrust. Everywhere men want to negotiate but only on their own special terms. We become major champions of minor issues because we have cut off too many options in dealing with the major issues. We may acknowledge our mutual fear, but we do not know what to do about it.

If persons involved in international conversations knew the language of P-A-C, if they could share the knowledge that the fear is in the Child, that there is no way to agree through the Parent, and only through the emancipation of the Adult can the universal I'M NOT OK—YOU'RE OK position be overcome, then we might begin to see possibilities of solutions beyond the limiting influences of the past. The basic words of Transactional Analysis (Parent, Adult, Child, NOT OK, OK, games, and stroking) are so simple that even if it were not possible to translate them into all other languages, they could be used as they are, with definitions in the words of those languages. "OK" is already an international word. Parent, Adult, and Child could also become international words. Now that we have a concept for understanding human behavior that all persons can comprehend, one which can be put into simple words and translated into any language, we may be arriving at a point where we can discard our archaic fears, based on the tragedies of the past, and begin talking with one another in the only way agreement on anything will be possible: Adult to Adult. With the Adult we can look together at some of the age-old hang-ups. Unexamined phrases close our options and our hopes for living together on an I'M OK—YOU'RE OK basis. For instance, how far can we get in world diplomacy if we continue to use a closed language with such neatly sealed phrases as "godless communism," the "free West," and "irreconcilable conflict"? Even the phrase "world communism," which conjures up such horrors that we are willing to continue to fight war after war at fantastic cost, is due for examination. How many wars will we have to fight? Is there an end in sight? Is world communism possible? Are all communists godless? What is a communist? Has he changed in the past fifty years? Are all communists alike?

There are three billion persons in the world. We know very little of these persons as individuals. We scarcely think of them as individuals. For instance, do we see a country like India only as a vast, nondescript blur of many-many-people whose importance is only in the way it shifts in the international balance of power in our fight against world communism? Or can we see India as a nation far more complicated, with real persons who make up one-seventh of the world's population, whose country contains six distinct ethnic groups, 845 languages and dialects, seven major religions, and two hostile cultures? If the Indian Parent and the American Parent cannot agree on anything, can we see the possibility of the excitement of the discovery of mutual concerns and shared joys through the emancipation of the Adult? We are related to one another, and we are persons and not things. The people of the world are not things to be manipulated, but persons to know; not heathen to be proselytized but persons to be heard; not enemies to be hated but persons to be encountered; not brothers to be kept but brothers to be brothers.

Impossible? Naïve? In an affluent society whose members are conditioned to believe that one man's problems cannot be resolved without taking the time of another man (trained three to five years in psychoanalysis after medical school and internship) for a period of thousands of hours over a period of several years, the thought of a solution to the needs of three billion people in crisis seems desperately absurd. The Parent says, "There will always be wars and rumors of wars." The Child says, "Eat, drink and be merry for tomorrow we die." History tells us what has been. But it cannot tell us what must be or cannot be. This is an open and evolving universe and we do not know enough about that universe to say what can't happen. Only the Adult can go to work on this exciting idea. Only the Adult has creative power.

The Adult can recognize the Child responses in others but can choose not to respond in kind. The United States, for instance, cannot have its way all the time. Robert Hutchins writes to this point in an article about the role of the United States:

> Let us admit the malevolence of China, the implacability of North Vietnam, the hostility of the Soviet Union, the eccentricity of De Gaulle and the instability of the undeveloped world. Let us remember at the same time that we live under the threat of thermo-nuclear incineration. What is the proper role of the United States in world affairs? What is the right and wise policy for it to follow?
>
> We are the victims not of the wickedness of others—that is a paranoid view—but of our own mistakes and delusions. This is not to deny that others are wicked. Of course they are. *What we have to do is avoid wickedness ourselves,* offer an example of magnanimous and intelligent power and organize the world to curb the inevitable wickedness we shall find at home and abroad.* [Italics mine.]

*R. Hutchins, article, San Francisco *Chronicle,* July 31, 1966.

The American myth seems to me to be grounded in the WE'RE OK—YOU'RE NOT OK position. WE'RE OK by virtue of our sentimental recollections of Patrick Henry and Thomas Jefferson and Thomas Paine and Abraham Lincoln. We think of ourselves in our best images but caricature the opposition. Thomas Merton in writing about today's angry world asks:

> What will we do when we finally have to realize that we are locked out of the lone prairie and thrust forth into a world of history along with all the wops and dagos and polacks: that we are just as much a part of history as all the rest of them? That is the end of the American myth: We can no longer lean out from a higher and rarefied atmosphere and point down from the firmament to the men on earth to show them the patterns of our ideal republic. We are in the same mess with all the rest of them. Shall we turn our backs on all that? Shall we open another can of beer and flip the switch and find our way back to the familiar mesquite, where all problems are easily solved; where the good guys are always the straight shooters and they always win?*

Shooting straight and winning are glorified in America by "good, honest, God-fearing people" who, slow of foot and dull of vision, wonder why violence now saturates the land. After the murder of Robert Kennedy, Arthur Miller wrote:

> There is violence because we have daily honored violence. Any half-educated man in a good suit can make his fortune by concocting a television show whose brutality is photographed in sufficiently monstrous detail. Who produces these shows, who pays to sponsor them, who is honored for acting in them? Are these people delinquent psychopaths slinking along tenement streets? No, they are the pillars of society, our honored men, our exemplars of success and social attainment. We must begin to feel the shame and contrition we have earned before we can

*T. Merton, *Conjectures of a Guilty Bystander* (New York: Doubleday, 1966).

begin to sensibly construct a peaceful society, let alone a peaceful world. A country where people cannot walk safely in their own streets has not earned the right to tell any other people how to govern itself, let alone to bomb and burn that people.*

The honoring of violence is recorded in the Parent of our little children. This gives permission to the rage and hatred present in the Child of any person. The combination is a death sentence for our culture. There are 6,500 murders per year committed in the United States compared with 30 in England, 99 in Canada, 68 in Germany, and 37 in Japan. More than two million guns were sold in the United States during 1967. During the first four months of 1968, in California alone, 74,241 pistols were bought—legally.

President Johnson charged his new crime study commission to look into the "causes, the occurrence, and the control of physical violence across this nation, from assassination that is motivated by prejudice and ideology and by politics and by insanity, to violence in our cities' streets and *even in our homes.*" [Italics mine.]

The violence in our homes is the most significant violence of all. It is the Child who commits murder. Where does the Child learn?

Every day one or two children under five years old in the United States are killed by their parents, according to Drs. Ray E. Helfer and C. Henry Kempe of the University of Colorado, who are reporting their findings in a book, *The Battered Child.* The infanticide rate is greater than the combined total taken by tuberculosis, whooping cough, polio, measles, diabetes, rheumatic fever, and appendicitis. In addition, every hour five infants are injured by their parents or guardian.

One of the problems involved in trying to find a solution to the problem, Dr. Heifer said, is finding psychiatrists to treat the parents. Help for parents was cited as a solution in a Gallup poll con-

*A. Miller, "The Trouble with Our Country," an article written for *The New York Times,* reprinted in the San Francisco *Chronicle,* June 16, 1968, p. 2.

ducted the day Senator Kennedy was shot. The responses indicated the first solution to the violence problem would be "stricter gun laws," but in addition a majority of those polled wanted: "stricter law enforcement . . . remove programs of violence from TV . . . improve parental control (including courses for parents on how to rear their children)."*

The **Institute for Transactional Analysis** in Sacramento has offered such courses since 1966. Several hundred parents have attended. The eight-week courses have begun with an explanation of P-A-C. The faculty has included psychiatrists, probation officers, ministers, pediatricians, educators, psychologists, and an obstetrician, all using the same language, P-A-C. Transactional Analysis has been applied to the following subjects: the dilemma "I Want to Trust Him, but . . ."; producing change in the youth offender; computing moral values; the relationship between freedom and love; problems of underachieving and handicapping; crisis intervention; why children "play stupid"; rescue and repair of "poor students"; building healthy attitudes toward sex and marriage; and emotional control. These courses have helped good parents become better parents and have helped to rescue and repair troubled families.

After completing one of these courses a mother wrote the following: "This course has been the best thing that has ever happened to us. It has opened a new line of communication between my husband and myself, and I feel that I personally benefited from it greatly. People I work with have constantly told me what a different person I have become since starting the course. One woman always tells me, 'God bless your Adult.' We also see what our problem is with our daughter and feel that we can certainly work it out ourselves."

Knowing how to stop violence in the home is knowing how to stop it in society. Our captains of industry, our advertisers, and our

*San Francisco *Chronicle*, June 16,1968.

producers of entertainment must learn the same thing these parents learn. Efforts in the home are undermined by the persistent input of contradictory data from without. My ten-year-old daughter asked if "we could go see *Bonnie and Clyde.*" I said no, it was full of violence and I did not like the way it glamorized some very sordid individuals. It was somewhat hard to explain a few days later why *Bonnie and Clyde* was repeatedly mentioned during the Academy Awards.

I believe people who have capitalized on violence have taken comfort from the point of view certain psychologists have held that watching violence is a safety valve which helps persons drain off violence rather than act it out. There is no way to validate this point of view. I believe there is mounting evidence, in fact, to invalidate it. These psychologists hold the view that feelings accumulate as if in a pail that every so often must be emptied. It is more accurate to think of feelings as a replay of old recordings which can be turned off at will. We do not have to go around dumping our feelings; we can simply turn them off, keep them from flooding our computer, and can, instead, fill that computer with something else. Emerson said, "A man is what he thinks about all day long."

In another age, when the world was filled with political murder, the selling of people into slavery, the crucifixion of innocent men, the killing of infants, and the cheers of entertainment moguls reveling over the blood in the arena, a wise and good man wrote to a little group of people in Philippi: "And now, my friends, all that is true, all that is noble, all that is just and pure, all that is lovable and gracious, whatever is excellent and admirable—fill all your thoughts with these things."*

We can hate evil so much that we forget to love good. And there is much that is good in America, much that in years past has inspired the admiration of people throughout the world, much that has drawn the oppressed from other shores. In 1950 Charles

*Paul's letter to the Philippians, 4:8.

Malik, then the Lebanese Ambassador to the United Nations, said:

> When I think of what your [America's] churches and univer-
> sities can do, by way of mediating love and forgiveness, impart-
> ing self-restraint, training the mind, revealing the truth, when I
> observe what your industries can accomplish by way of trans-
> forming this whole material universe into an instrument which
> will lighten the burden of man; when I ponder what your homes
> and small communities can create, by way of character and so-
> lidity and stability and humor; and when I reflect on the great
> media of the newspaper, the cinema, the radio and television, and
> how they can immensely help in the articulation of the American
> word; when I humbly and concretely think on these things, and
> when I further meditate that there is nothing to prevent all these
> agencies from dedicating themselves to truth and love and being;
> then I say, perhaps the day of the Lord is at hand.

The only thing that may prevent this dedication is fear—fear of other persons on this earth, *fear in the Child,* which will drain our resources for good into an ever-escalating battle that we mistakenly think we can win.

Winners and Losers

Hamlet's alternatives were "to be or not to be." Our national alternatives are believed to be "to win or not to win" the struggle against
world communism. To win is more important than to be, it would
appear, in view of the increasing risk of the final armed aggression
that will lead to global incineration. A Vietnam village is shelled so
thoroughly that when troops finally enter it there is nothing standing and no one living. The commander of an operation of this sort
is quoted as saying, "We had to destroy them to save them." This
sounds very much like the Parent pronouncement which came *ex*

cathedra from the woodshed: This hurts me more than it hurts you. Can we really say to the destroyed village littered with its charred inhabitants, This hurts us more than it hurts you?

How do the people of Asia really see the democracy we extol and insist is best for them? Do they like it? Do they understand it? Do they judge "our free way of life" by what they see going on in our country? Do they believe we really love the non-Caucasian Asians in view of the racial strife in our country? We say, "Democracy is delicious," in the same way that Mother said, "Spinach is delicious." We were not allowed to report the evidence of our taste buds. In many similar transactions we were forced to mistrust our own senses and not acknowledge our own emotions. Was Mother really that enthusiastic about spinach herself? How enthusiastic are we about our democratic institutions? Democracy *is* a good thing, but is violence and war the only way we can establish its goodness?

"Democracy is delicious" and "This hurts me more than it hurts you" are both extremely dangerous international games in that they are *ulterior* to the real motive, which is: "We must win, for if we do not win, then we lose."

Are winning and losing the only options for persons or for nations? The only way to stay a winner is to surround oneself with losers. Winners and losers have been the only models we have had. When the primates were driven from the forests by the climatic changes that reduced the size of the forests, there were only two possible outcomes to their encounter with the old-time carnivores on the open plain. Those who won the battle over food survived; those who lost died. It is true that religious and political leaders have emerged from time to time with what they claimed to be a new model; yet, for the most, the ideas of these "dreamers and prophets" have been discounted as Utopian, otherworldly, and impossible. The fact is that the winning and losing models have predominated throughout the history of mankind.

But circumstances have changed. Because of scientific knowledge enough food can be produced to feed the people of the world if the population explosion can be halted. Science and education have also made birth control possible. It is now possible to conceive of another option: I'M OK—YOU'RE OK. Coexistence is at last a possibility based on reality. In the beginning, man grew and developed in service to his own survival. Can we now turn to new tasks, to the tasks of the survival of all the people of the world? Can the gift of life and our brief span of existence on this earth be enjoyed to the fullest of human spiritual capabilities?

If we see that I'M OK—YOU'RE OK is at last within the realm of possibility, do we dare look for change, something new under the sun, something to stop the violence threatening to destroy what has taken millions of years to build?

Teilhard stated: "Either nature is closed to our demands for futurity, in which case thought, the fruit of millions of years of effort, is stifled, stillborn, in a self-abortive and absurd universe. Or else an opening exists. . . ."*

We believe we have found an opening. This opening will be explored not by a nameless, corporate society but by individuals together in that society. The exploration can be made only as individuals are emancipated from the past and become free to choose either to accept or reject the values and methods of the past. One conclusion is unavoidable: Society cannot change until persons change. We base our hope for the future on the fact that we have seen persons change. How they have done it is the good news of this book. We trust it may be a volume of hope and an important page of the manual for the survival of mankind.

*Pierre Teilhard de Chardin, *The Phenomenon of Man* (New York: Harper, 1961).

Index

Made in the USA
San Bernardino, CA
06 May 2019